FLECK

**Books are to be returned on
the last date below.**

FLEC. LIBRARY

Reaction Mechanisms of
Inorganic and Organometallic
Systems

Reaction Mechanisms of Inorganic and Organometallic Systems

ROBERT B. JORDAN

New York   Oxford
OXFORD UNIVERSITY PRESS
1991

Oxford University Press

Oxford New York Toronto
Delhi Bombay Calcutta Madras Karachi
Petaling Jaya Singapore Hong Kong Tokyo
Nairobi Dar es Salaam Cape Town
Melbourne Auckland

and associated companies in
Berlin Ibadan

Library of Congress Cataloging-in-Publication Data
Jordan, Robert B.
Reaction mechanisms of inorganic and organometallic systems /
Robert B. Jordan.
p. cm. Includes bibliographical references and index.
ISBN 0-19-506945-5
1. Chemical reaction, Conditions and laws of. 2. Organometallic compounds.
3. Inorganic compounds. I. Title.
QD502.J67 1991
541.3'9—dc20 90-27512

Preface

This book has evolved from the lecture notes of the author for a one-semester course given to senior undergraduates and graduate students over the past 20 years. There is more material than can be covered in depth in one semester, but the organization allows the lecturer to give less coverage to certain areas without jeopardizing an understanding of other areas. It is assumed that the students are familiar with elementary crystal field theory and its applications to electronic spectroscopy and energetics, and concepts of organometallic chemistry, such as the 18-electron rule, π bonding, and coordinative unsaturation. The terminology and developments of elementary kinetics are given in the first two chapters, where some background from a physical chemistry course would be useful, and a familiarity with simple differential and integral calculus is assumed.

The material generally has been updated to developments through 1990, with references to the original literature provided along with recent review articles. It is expected that students will obtain further information from these sources and gain a feeling for the excitement in the field and an ability to evaluate such work critically. Some sample problems covering the work in each chapter are given at the end of the book.

The issue of units continues to be a vexing one in this area. A major goal of this course has been to provide students with sufficient background that they can read and analyze current research papers. To do this and be able to compare results, the reader must be ever vigilant about the units used by different authors. There is now general agreement that the time unit for rate constants is s^{-1} and this has been adopted throughout. Energy units are a different matter. Since both joules and calories are in common usage, both units have been retained in the text, with the choice made on the basis of the units in the original work as much as possible. However, within individual sections the text attempts to use one energy unit. In spectroscopic areas the cm^{-1} is dominant, but spectra are often given in nanometers and the conversion between these units and calories and joules is given when it seems important to understanding the issue being covered. Bond lengths are given in angstroms, which are still commonly used in crystallography, and in picometers, which are used in theoretical calculations. The formulas for such calculations are given in the original or most common format and units for the various quantities are always specified.

The author is indebted to all those whose research efforts have provided the core of material for this book. The author is pleased to acknowledge those who have provided the inspiration for this book: first, my parents, who contributed the early atmosphere and encouragement; second, Henry Taube, whose intellectual and experimental guidance ensured my continuing enthusiasm for mechanistic studies. Finally and foremost, Anna has been a vital force in the creation of this book through her understanding of the time commitment, her force of will to see the project completed, and her invaluable comments, criticisms, and assistance in producing the manuscript.

R. B. J.

Edmonton, Alberta
March 1991

Contents

Reaction Mechanisms of
Inorganic and Organometallic
Systems

1

Tools of the Trade

This chapter covers the basic terminology related to the types of studies that are commonly used to provide information about a reaction mechanism. More background material is available from physical chemistry texts.[1,2] Experimental techniques are not discussed in this book, but such information is available from the recent review by Wilkins[3] on rapid reaction methods and from the earlier book by Caldin.[4] The reader is also referred to the initial volumes of the series edited by Bamford and Tipper.[5]

1.1 BASIC TERMINOLOGY

As with most fields, the study of reaction kinetics has some terminology with which one must be familiar in order to understand advanced books and research papers in the area. The following is a summary of some of these basic terms and definitions. Many of these should be familiar from previous studies in introductory and physical chemistry, and further background can be obtained from textbooks devoted to the physical chemistry aspects of reaction kinetics.

Rate Law
The rate law is the experimentally determined dependence of the reaction rate on reagent concentrations. It has the following general form:

$$\text{Rate} = k\,[A]^m\,[B]^n \ldots \tag{1.1}$$

where k is a proportionality constant called the rate constant. The exponents m and n are determined experimentally from the kinetic study. It should be noted that the exponents in the rate law have no necessary relationship to the stoichiometric coefficients in the balanced chemical reaction, and the rate law may even contain species that do not appear in the balanced reaction.

The rate law is an essential piece of mechanistic information because it contains the concentrations of species necessary to get from the reactant to the product by the lowest energy pathway. A fundamental requirement of an acceptable mechanism is that it must predict a rate law consistent with the experimental rate law.

Order of the Rate Law

The order of the rate law is the sum of the exponents in the rate law. For example, if m = 1 and n = –2 in Eq. (1.1), the rate law has an overall order of –1. However, except in the simplest cases, it is best to describe the order with respect to individual reagents; in this example, first order in [A] and inverse second order in [B].

Rate Constant

The rate constant (k) is the proportionality constant that relates the rate to the reagent concentrations (or activities or pressures, for example), as shown in Eq. (1.1). The units of k depend on the rate law and must give the right-hand side of Eq. (1.1) the same units as the left-hand side.

Half-time

The half-time $(t_{1/2})$ is the time required for a reactant concentration to change by half of its total change. This term is used to convey a qualitative idea of the time scale and has a quantitative relationship to the rate constant in simple cases. In complex systems, the half-time may be different for different reagents and one should specify the reagent to which the $t_{1/2}$ refers.

Lifetime

For a particular species, the lifetime (τ) is the concentration of that species divided by its rate of disappearance. This term is commonly used in so-called lifetime methods, such as nmr, and in relaxation methods, such as temperature-jump.

1.2 ANALYSIS OF RATE DATA

In general, a kinetic study begins with the collection of data of concentration versus time of a reactant or product. As will be seen later, this can also be accomplished by determining the time dependence of some variable that is proportional to concentration, such as absorbance or nmr peak intensity. The next step is to fit the concentration–time data to some model that will allow one to determine the rate constant if the data fits the model.

The following section develops some integrated rate laws for the models most commonly encountered in inorganic kinetics. This is essentially a mathematical problem; given a particular rate law as a differential equation, the equation must be reduced to one concentration variable and then integrated. The integration can be done by standard methods or by reference to integration tables. Many more complex examples are given in advanced textbooks on kinetics.

1.2.a Zero-Order Reaction

A zero-order reaction is rare for inorganic reactions in solution but is included for completeness. For the general reaction

$$A \xrightarrow{k} B \tag{1.2}$$

the zero-order rate law is given by

$$\frac{d[B]}{dt} = k \tag{1.3}$$

and integration over the limits $[B] = [B]_0$ to $[B]$ and $t = 0$ to t yields

$$[B] - [B]_0 = kt \tag{1.4}$$

This predicts that a plot of $[B]$ or $[B] - [B]_0$ versus t should be linear with a slope of k.

1.2.b First-Order System Coming to Equilibrium

Normally, the simple first-order system would be described at this point but, since it can be generated as a special case of the equilibrium situation, we will start with the more complex system described by Eq. (1.5):

$$A \underset{k_{-1}}{\overset{k_1}{\rightleftharpoons}} B \tag{1.5}$$

The rate of disappearance of A equals the rate of appearance of B and

$$-\frac{d[A]}{dt} = \frac{d[B]}{dt} = k_1[A] - k_{-1}[B] \tag{1.6}$$

The problem is to convert this equation to a form with only one concentration variable, either $[A]$ or $[B]$, and then integrate the equation to obtain the *integrated rate law*. The choice of the variable to retain will depend on what has actually been measured experimentally. The elimination of one concentration is done by considering the reaction stoichiometry and the initial conditions. The most general conditions are that both A and B are present initially at concentrations $[A]_0$ and $[B]_0$, respectively. If the concentrations are defined as $[A]$ and $[B]$ at any time and as $[A]_e$ and $[B]_e$ at equilibrium, then from mass balance

$$[A]_0 + [B]_0 = [A] + [B] = [A]_e + [B]_e \tag{1.7}$$

therefore

$$[A] = [A]_0 + [B]_0 - [B] \quad \text{and} \quad [A]_0 = [A]_e + [B]_e - [B]_0 \tag{1.8}$$

so that

$$[A] = [A]_e + [B]_e - [B]_0 + [B]_0 - [B] = [A]_e + [B]_e - [B] \qquad (1.9)$$

Now one can substitute for [A] from Eq. (1.9) into Eq. (1.6) to obtain

$$\frac{d[B]}{dt} = k_1 ([A]_e + [B]_e) - (k_1 + k_{-1}) [B] \qquad (1.10)$$

Note that the initial concentrations have been eliminated.

Since the preceding equation contains only one concentration variable, [B], it can be integrated directly. However, it is convenient in the end to eliminate $[A]_e$ by noting that, at equilibrium, the rate in the forward direction must be equal to the rate in the reverse direction:

$$k_1 [A]_e = k_{-1} [B]_e \qquad (1.11)$$

and substitution for $k_1[A]_e$ into Eq. (1.10) gives

$$\frac{d[B]}{dt} = k_{-1} [B]_e + k_1 [B]_e - (k_1 + k_{-1}) [B]$$

$$= (k_1 + k_{-1})([B]_e - [B]) \qquad (1.12)$$

This equation can be rearranged and integrated over the limits $[B] = [B]_0$ to [B] and t = 0 to t to obtain

$$\ln([B]_e - [B]) - \ln([B]_e - [B]_0) = - (k_1 + k_{-1}) t \qquad (1.13)$$

Therefore, a plot of $\ln([B]_e - [B])$ versus t should be linear with a slope of $- (k_1 + k_{-1})$. Note that the kinetic study yields the sum of the forward and reverse rate constants. If the equilibrium constant (K) is known, then k_1 and k_{-1} can be calculated since $K = k_1 / k_{-1}$.

A very important practical advantage of the first-order system is that *the analysis can be done without any need to know the initial concentrations.* This means that the collection of concentration–time data can be started at any time arbitrarily defined as t = 0.

At the reaction half-time, $t = t_{1/2}$, $[B] = 1/2([B]_e - [B]_0) + [B]_0$, and substitution into Eq. (1.13) gives

$$t_{1/2} = \frac{\ln 2}{k_1 + k_{-1}} = \frac{0.693}{k_1 + k_{-1}} \qquad (1.14)$$

Therefore, *the half-time is independent of the initial concentrations.*

1.2.c First-Order Irreversible System

Strictly speaking, there is no such thing as an irreversible reaction. It is just a system in which the rate constant in the forward direction is much larger than that in the reverse direction. Since $K = k_1/k_{-1}$, this corresponds to having a very large equilibrium constant for the reaction, so that Eq. (1.5) simplifies to

$$A \xrightarrow{\quad k_1 \quad} B \qquad (1.15)$$

The kinetic analysis of the irreversible system is a special case of the reversible system just described in which $k_1 \gg k_{-1}$. Then $[B]_e$ is replaced by $[B]_\infty$, the final concentration of B at "infinite time." In addition, mass balance requires that

$$[B]_\infty = [A]_0 + [B]_0 \qquad (1.16)$$

so that substitution into Eq. (1.13), noting that $(k_1 + k_{-1}) = k_1$ because $k_1 \gg k_{-1}$, gives

$$\ln([B]_\infty - [B]) - \ln([B]_\infty - [B]_0) = -k_1 t \qquad (1.17)$$

A plot of $\ln([B]_\infty - [B])$ versus t should be linear with a slope of $-k_1$.

At the reaction half-time, $[B] = 1/2([B]_\infty - [B]_0) + [B]_0$, and substitution into Eq. (1.17) gives

$$t_{1/2} = \frac{\ln 2}{k_1} = \frac{0.693}{k_1} \qquad (1.18)$$

A more familiar form of the integrated rate law in terms of A can be obtained by substitution from Eq. (1.16) into Eq. (1.17) and noting that

$$[B] = [B]_0 + [A]_0 - [A] \qquad (1.19)$$

to give

$$-\ln[A] + \ln[A]_0 = k_1 t \qquad (1.20)$$

or

$$[A] = [A]_0 e^{-k_1 t} \qquad (1.21)$$

The same form is obtained by directly integrating $-d[A]/dt = k_1[A]$.

The first-order irreversible solution has the same mathematical form as the reversible case and the same advantages with regard to initial conditions.

1.2.d   Second-Order System Coming to Equilibrium

This system can be described by

$$A + B \; \underset{k_{-2}}{\overset{k_2}{\rightleftarrows}} \; C \tag{1.22}$$

and the rate of formation of C is given by

$$\frac{d[C]}{dt} = k_2 [A][B] - k_{-2}[C] \tag{1.23}$$

To simplify the development, one can assume that there is no C present initially and that the stoichiometry is $1 : 1 : 1$, so that mass balance gives initially

$$[A] = [A]_0 - [C] \quad \text{and} \quad [B] = [B]_0 - [C] \tag{1.24}$$

and at equilibrium

$$[A]_e = [A]_0 - [C]_e \quad \text{and} \quad [B]_e = [B]_0 - [C]_e \tag{1.25}$$

If the stoichiometry is other than $1 : 1 : 1$, then appropriate stoichiometry coefficients must be used in the mass balance conditions and in Eq. (1.23).

Substitution of the initial conditions into Eq. (1.23) gives an equation that can be integrated because [C] is the only concentration variable. However, it proves convenient to eliminate k_{-2} from the equation before integrating by noting that the forward and reverse rates are equal at equilibrium:

$$k_2 ([A]_0 - [C]_e)([B]_0 - [C]_e) = k_{-2} [C]_e \tag{1.26}$$

so that

$$k_{-2} = \frac{k_2 ([A]_0 - [C]_e)([B]_0 - [C]_e)}{[C]_e} \tag{1.27}$$

and substitution into Eq. (1.23) yields

$$\frac{d[C]}{dt} = k_2 \frac{([C]_e - [C])([A]_0 [B]_0 - [C]_e [C])}{[C]_e} \tag{1.28}$$

This equation can be rearranged and integrated over the limits $[C] = 0$ to $[C]$ and $t = 0$ to t to give the following solution:

$$\ln\left(\frac{[A]_0\,[B]_0 - [C]_e\,[C]}{[C]_e\left([C]_e - [C]\right)}\right) + \ln\left(\frac{[C]_e^2}{[A]_0\,[B]_0}\right) = \left(\frac{\left([A]_0\,[B]_0 - [C]_e^2\right)k_2}{[C]_e}\right)t \quad (1.29)$$

A plot of the first term on the left-hand side of Eq. (1.29) versus t should be linear with a slope related to k_2, as indicated by the right-hand side of Eq. (1.29). It is apparent that *one must know the initial concentrations, $[A]_0$ and $[B]_0$, and the final concentration, $[C]_e$, in order to do the analysis and to determine the value of k_2 from the slope*. These requirements make this an unpopular and uncommon situation for experimental studies.

1.2.e   Second-Order Irreversible System

This system can be obtained as a special case of the reversible system by simple consideration of the stoichiometry conditions. If $[A]_0 < [B]_0$ and the reaction in Eq. (1.22) goes essentially to completion, then $[C]_e = [A]_0$, and substitution of this condition into Eq. (1.29) gives

$$\ln\left(\frac{[B]_0 - [C]}{[A]_0 - [C]}\right) + \ln\left(\frac{[A]_0}{[B]_0}\right) = \left([B]_0 - [A]_0\right)k_2 t \quad (1.30)$$

In this case, the initial concentrations of both reactants are required in order to plot the first term on the left versus t and to determine k_2 from the slope. These conditions are not as restrictive as those for the reversible second-order system, but they are still worse than those for the first-order system.

At the half-time for this second-order reaction, $[C] = [A]_0/2$, and substitution into Eq. (1.30) shows that

$$t_{1/2} = \frac{1}{\left([B]_0 - [A]_0\right)k_2}\ln\left(\frac{2\,[B]_0 - [A]_0}{[B]_0}\right) \quad (1.31)$$

1.2.f   Pseudo-First-Order Reaction Conditions

The pseudo-first-order reaction condition is very widely used, but it is seldom mentioned in textbooks. Although many reactions have second-order or more complex rate laws, the experimental kineticist wishes to optimize experiments by taking advantage of the first-order rate law since it imposes the fewest restrictions on the conditions required to determine a reliable rate constant. The trick is to use the pseudo-first-order condition.

The *pseudo-first-order condition* is that the concentration of the reactant whose concentration is monitored is much smaller (<10 times) than that of all the other reactants, so that the concentrations of all the latter remain essentially constant during the reaction. Under this condition, the rate law usually simplifies to a first-order form and one gains the advantage of not needing to know the initial concentration of the deficient reagent.

In the preceding irreversible second-order example, if it is assumed that the conditions have been set so that $[B]_0 \gg [A]_0$, then $[B]_0 \gg [C]$. In addition, the concentration of B will remain constant at $[B]_0$, and the final concentration of C is $[C]_\infty = [A]_0$ if the reaction is irreversible and has 1 : 1 stoichiometry. Substitution of these conditions into Eq. (1.30) gives

$$\ln\left(\frac{[C]_\infty}{[C]_\infty - [C]}\right) = [B]_0 k_2 t \tag{1.32}$$

This equation predicts that a plot of $\ln([C]_\infty - [C])$ versus t should be linear with a slope of $-k_2[B]_0$. This is identical in form to the first-order rate law except that k_1 is replaced by $k_2[B]_0$. The latter constant is often called the observed or experimental rate constant (k_{obsd} or k_{exp}). Since $[B]_0$ is known, it is possible to calculate k_2.

In a more general case, if the rate of disappearance of reactant A is given by

$$-\frac{d[A]}{dt} = k[X]^x[Y]^y[Z]^z[A] \tag{1.33}$$

and the conditions are such that $[X]_0, [Y]_0, [Z]_0 \gg [A]$, then

$$-\frac{d[A]}{dt} = k[X]_0^x[Y]_0^y[Z]_0^z[A] = k_{exp}[A] \tag{1.34}$$

and the rate law has the first-order form.

1.2.g   Comparison of First-Order and Second-Order Conditions

It is helpful to compare the time dependence of the formation of C for the first- and second-order conditions for the case where the rate constant is the same and the initial concentrations are changed.

In Figure 1.1, $[A]_0 = 0.10$ M and $k_2 = 3 \times 10^{-3}$ M^{-1} s^{-1}. The heavy line shows a pseudo-first-order dependence of [C] when $[B]_0 = 10 \times [A]_0 = 1.0$ M, so that $k = [B]_0 k_2 = 3 \times 10^{-3}$ s^{-1}. Under second-order conditions, such as $[B]_0 = 5 \times [A]_0 = 0.50$ M, the time dependence of [C] follows a second-order rate law and [C] increases more slowly than in the pseudo-first-order case. As $[B]_0$ increases to 0.75 M, the second-order curve becomes more similar to the first-order curve, and when $[B]_0 = 10 \times [A]_0$, the two curves are nearly indistinguishable. The system is now approaching a pseudo-first-order condition; this is the rationale for the general rule that pseudo-first-order conditions require at least a 10-fold excess of the reagents whose concentrations are to remain constant.

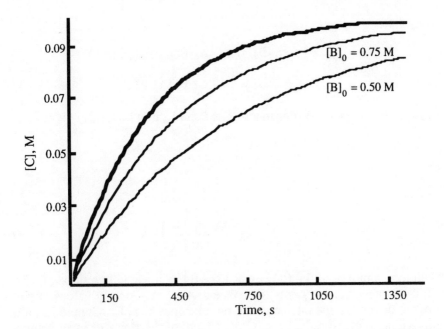

Figure 1.1. The time dependence of [C] for first-order (—) and various second-order (—) conditions with $[A]_0 = 0.10$ M.

1.3 CONCENTRATION VARIABLES AND FIRST-ORDER RATES

The rate laws have been developed in terms of concentrations, but in many cases it is not practical or possible to determine actual molar concentrations as a function of time. However, it is easy to measure some property that is known to be directly proportional to molar concentration, such as absorbance, nmr or ir integrated peak intensity, conductance, or refractive index. In such cases, the first-order system can still be analyzed to determine the rate constant.

The following development assumes an irreversible system but can easily be expanded to the more general reversible case. The system and integrated rate law are given by Eqs. (1.15) and (1.21), respectively. If the property being observed is called I and the proportionality constant with concentration is ε, then initially

$$I_0 = \varepsilon_A [A]_0 \qquad (1.35)$$

at the end

$$I_\infty = \varepsilon_B [B]_\infty = \varepsilon_B [A]_0 \qquad (1.36)$$

at any time

$$I = \varepsilon_A [A] + \varepsilon_B [B] = \varepsilon_A [A] + \varepsilon_B ([A]_0 - [A])$$

$$= (\varepsilon_A - \varepsilon_B)[A] + I_\infty \qquad (1.37)$$

Substitution for [A] and $[A]_0$ in terms of I and I_∞ into Eq. (1.21) gives

$$\frac{(I - I_\infty)}{(\varepsilon_A - \varepsilon_B)} = \frac{I_\infty}{\varepsilon_B} e^{-k_1 t} \qquad (1.38)$$

so that

$$\ln(I - I_\infty) = \ln\left(\frac{(\varepsilon_A - \varepsilon_B) I_\infty}{\varepsilon_B}\right) - k_1 t \qquad (1.39)$$

Therefore, a plot of $\ln(I - I_\infty)$ versus t should be linear with a slope of $-k_1$.

It is not necessary to know the concentration of A or any values of ε in order to determine the rate constant, but one does need I_∞. Sometimes it is impossible to measure I_∞ because of secondary reactions, or inconvenient because the reaction is slow. Such systems can be analyzed by nonlinear least-squares fitting of the data over as much of the reaction as possible, or in a more classical way by the Guggenheim method described in more detail by Moore and Pearson[1] (p. 71), Mangelsdorf,[6] and Espenson[7] (p. 25).

1.4 COMPLEX RATE LAWS

It is not unusual for a rate law to be more complex than the simple zero-, first-, or second-order cases we have considered. In general, the rate law has the following form:

$$-\frac{d[A]}{dt} = f([X], [Y], [Z])[A] = k_{exp}[A] \qquad (1.40)$$

where $f([X], [Y], [Z])$ is a function of the concentrations of X, Y, and Z.

In inorganic systems, some common forms of k_{exp} are

$$k_{exp} = k'[X] \qquad\qquad k_{exp} = k' + k''[X]$$

$$k_{exp} = \frac{k'[X]}{k'' + [Y]} \qquad\qquad k_{exp} = \frac{k'[Y] + k''}{k''' + [Z]} \qquad (1.41)$$

The dependence of k_{exp} on [X], [Y], and [Z] is determined from a series of kinetic experiments under pseudo-first-order conditions, keeping [Y] and [Z] constant and changing [X] to determine the dependence of k_{exp} on [X], and then repeating the process for Y and Z.

In many elementary considerations of kinetic data, it is suggested that the order of a reaction with respect to a particular reagent be determined from a plot of the logarithm of the rate constant versus the logarithm of the reagent concentration. This procedure is only appropriate for simple forms such as the first example in Eq. (1.41). Although log–log plots may appear linear for the more complex forms, the plots will yield meaningless fractional orders and should be avoided. Unfortunately, there is no truly general method of analysis to yield the reaction order, but this is seldom a serious problem when the reagent concentrations have been varied over a reasonable concentration range.

1.5 COMPLEX KINETIC SYSTEMS

Sometimes, even under pseudo-first-order conditions, the kinetic observations do not obey the first-order integrated rate law. This may indicate a number of chemical problems, such as impurities, a nonlinear analytical method, or precipitate formation. However, it is also possible that the system is more complex, with parallel and/or successive reactions, as shown in Eq. (1.42):

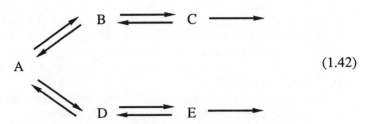

$$(1.42)$$

If pseudo-first-order conditions are maintained, it is always possible to solve the differential equations to determine the integrated rate law.[8,9] The solution has the general form

$$[\text{Product}] = M\, e^{-\gamma_1 t} + N\, e^{-\gamma_2 t} + \cdots \qquad (1.43)$$

where M, N, ... and γ_1, γ_2, ... are constants that depend on the rate constants for the individual steps. The number of exponential terms equals the number of steps in the reaction network. For such systems, the time dependence of the concentration variable is usually fitted by nonlinear least squares to determine the constants. In practice, exceptionally good data are required to extract more than two γ values.

A system encountered with some frequency is described by

$$A \underset{\beta_1}{\overset{k_1}{\rightleftharpoons}} B \underset{\beta_2}{\overset{k_2}{\rightleftharpoons}} C \qquad (1.44)$$

If one measures some property I_t that is directly proportional to concentration, then the integrated rate law predicts the time dependence for I_t given by

$$I_t = I_\infty + \left(\frac{1}{\gamma_2 - \gamma_1}\right) \times$$

$$\left\{\left[k_1(I_B - I_A) - \gamma_2(I_\infty - I_A)\right]e^{-\gamma_1 t} - \left[k_1(I_B - I_A) - \gamma_1(I_\infty - I_A)\right]e^{-\gamma_2 t}\right\} \quad (1.45)$$

where I_A and I_B are the values of the property I for species A and B, respectively, I_∞ is the final value of I at "infinite time," and γ_1 and γ_2 are the apparent rate constants. The latter are related to the specific rate constants in Eq. (1.44) by

$$\gamma_{1,2} = \frac{k_1 + \beta_1 + k_2 + \beta_2}{2} \pm$$

$$\frac{\sqrt{(k_1 + \beta_1 + k_2 + \beta_2)^2 - 4(k_1 k_2 + k_1 \beta_2 + \beta_1 \beta_2)}}{2} \qquad (1.46)$$

The form of Eq. (1.45) is useful for computer fitting procedures because one usually has some idea of reasonable values of I_A and I_B and an experimental value for I_∞. The equation for simpler schemes in which either or both of the reverse rate constants are zero can be obtained by setting the appropriate terms equal to zero in Eq. (1.46).

A somewhat simpler example of this type is a system that proceeds by several parallel paths to give different products, as shown in Eq. (1.47):

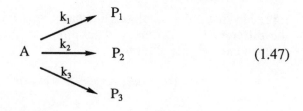

$$(1.47)$$

The rate and integrated rate law are given by

$$-\frac{d[A]}{dt} = (k_1 + k_2 + k_3)[A] \qquad (1.48)$$

$$[A] = [A]_0 e^{-(k_1 + k_2 + k_3)t} \qquad (1.49)$$

Therefore, the kinetics will give $k_{exp} = (k_1 + k_2 + k_3)$, but it is necessary to determine the final product distribution in order to evaluate the individual k_i values from the relationship

$$[P_i]_\infty = \frac{k_i\,[A]_0}{k_1 + k_2 + k_3} \qquad (1.50)$$

Such systems are discussed in more detail by Espenson[7] (pp. 55–56).

1.6 TEMPERATURE DEPENDENCE OF RATE CONSTANTS

To obtain information about the energetics of a reaction, the temperature dependence of the rate constant is determined. For complex rate laws, this will also involve a study of the concentration dependence of the rate at different temperatures, in order to determine the temperature dependence of the various terms contributing to the rate law. Once the experimental information is available for the specific rate constants, it is usually analyzed in terms of one of the following formalisms.

1.6.a Arrhenius Equation

Arrhenius seems to have been the first to find empirically that rate constants have a temperature dependence analogous to that of equilibrium constants, as given by the following exponential or logarithmic forms:

$$k = A\exp\!\left(-\frac{E_a}{RT}\right) \qquad (1.51)$$

$$\ln k = \ln A - \frac{E_a}{R}\left(\frac{1}{T}\right) \qquad (1.52)$$

where A is called the Arrhenius pre-exponential factor, E_a is the Arrhenius activation energy, R is the gas constant, and T is the temperature in Kelvin. The units of A will be the same as those of k and E_a will be in cal mol^{-1} or J mol^{-1}, depending on the units chosen for R (1.987 cal mol^{-1} K^{-1} or 8.314 J mol^{-1} K^{-1}). The k is determined by appropriate experiments at several temperatures, and a plot of ln k versus T^{-1} has a slope of $-E_a/R$. The rate constant is always predicted to increase with increasing temperature. Typical values are 10 to 30 kcal mol^{-1} for E_a, and for a first-order rate constant, 10^{10} to 10^{14} s^{-1} for A.

The Arrhenius equation is still widely used in certain areas of kinetics and for complex systems where the measured rate constant is thought to be a complex composite of specific rate constants.

1.6.b   Transition-State Theory

This theory was developed originally for a simple dissociation process in the gas phase and it assumes that the reaction can be described by the following sequence:

$$A—B \; \underset{}{\overset{K^*}{\rightleftharpoons}} \; \{ A\text{----}B \}^* \; \xrightarrow{k_3} \; A + B \qquad (1.53)$$

Activated complex
or transition state

The theory proposes that the activated complex or transition state will proceed to products when the bond has thermal energy kT, so that its vibrational frequency is $v = kT/h \; s^{-1}$ ($k \equiv$ Boltzmann's constant, 1.38×10^{-16} erg K^{-1}; $h \equiv$ Planck's constant, 6.622×10^{-27} erg s), which will equal the rate constant k_3. Furthermore, it is assumed that the activated complex is always in equilibrium with the reactant, so that $K^* = [A\text{----}B]^* / [A—B]$. If K^* is a normal equilibrium constant, then

$$\frac{d[B]}{dt} = k_3 [A\text{----}B]^* = k_3 K^* [A—B] = \frac{kT}{h} K^* [A—B] \qquad (1.54)$$

and

$$\ln K^* = -\frac{\Delta G^{o*}}{RT} = -\frac{\Delta H^{o*}}{RT} + \frac{\Delta S^{o*}}{R} \qquad (1.55)$$

where ΔG^{o*}, ΔH^{o*}, and ΔS^{o*} are the standard molar free energy, enthalpy, and entropy differences, respectively, between the activated complex and the reactants.

If the first-order rate expression, $d[B]/dt = k_{exp} [A—B]$, is compared to Eq. (1.54), then substitution from Eq. (1.55) shows that

$$k_{exp} = \frac{kT}{h} K^* = \frac{kT}{h} \exp\left(-\frac{\Delta G^{o*}}{RT}\right) \qquad (1.56)$$

This expression can be rearranged, expanded by substitution from Eq. (1.55), and put into logarithmic form to give

$$\ln\left(\frac{k_{exp}}{T}\right) = \ln\left(\frac{k}{h}\right) - \frac{\Delta H^{o*}}{RT} + \frac{\Delta S^{o*}}{R} \qquad (1.57)$$

Therefore, a plot of $\ln(k_{exp}/T)$ versus T^{-1} should be linear with a slope of $-\Delta H^{o*}/R$. The value of ΔS^{o*} can then be calculated from a known value of k_{exp} at a particular T:

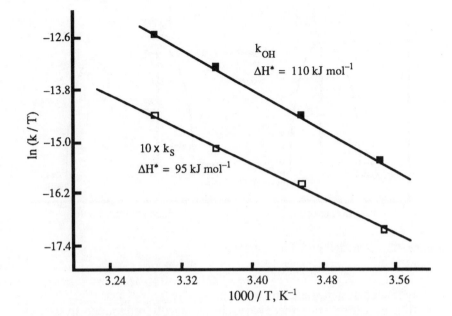

Figure 1.2. The temperature dependence of the isomerization of $(NH_3)_5Co(ONO)^{2+}$.

$$\Delta S^{o*} = 4.576\left[\log\left(\frac{k_{exp}}{T}\right) - 10.319\right] + \frac{\Delta H^{o*}}{T} \ (cal\ mol^{-1}\ K^{-1}) \quad (1.58)$$

In common usage, the standard state designations are dropped and ΔH^* and ΔS^* are called the activation enthalpy and activation entropy, respectively, for the reaction.

The Arrhenius parameters are related to ΔH^* and ΔS^* by the following relationships:

$$\Delta H^* = E_a - RT \quad \text{and} \quad \Delta S^* = 4.58\ (\log A - 13.2) \quad (1.59)$$

Some typical data[10] for the temperature dependence for the linkage isomerization of $(NH_3)_5Co(ONO)^{2+}$ to form $(NH_3)_5Co(NO_2)^{2+}$ are shown in Figure 1.2. The rate law for the reaction in alkaline solution under pseudo-first-order conditions shows a spontaneous path (k_S) and an OH^- catalyzed path (k_{OH}). The ΔH^* is somewhat higher for the k_{OH} path, as indicated by the steeper slope in Figure 1.2.

The transition-state theory parameters are used commonly to describe the temperature dependence of k_{exp} even for processes in solution that are far more complex than assumed in the original formulation of the theory. Therefore, ΔH^* and ΔS^* are best treated as experimental parameters that are useful for calculating k_{exp} at other temperatures and for comparison to other

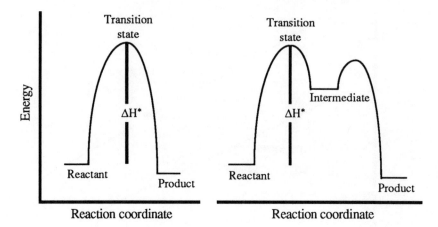

Figure 1.3. Reaction coordinate diagrams.

closely related systems and processes.

Within transition-state theory, reactions are often depicted in terms of "reaction coordinate diagrams." These are plots of the energy of the system versus the "reaction coordinate," which is an ambiguous measure of the extent to which reactant has been converted to product. Examples of such diagrams are shown in Figure 1.3.

The *transition state* (or activated complex) is the species at the highest energy point on the reaction coordinate diagram. An *intermediate* is the species present at any valley on a reaction coordinate diagram. When a valley is shallow, it can be ambiguous whether or not there is an intermediate really formed. In chemistry, an intermediate is expected to have a lifetime longer than a few vibrational lifetimes ($>10^{-13}$ s) and the valley should be deeper than the thermal energy (RT = 600 cal mol^{-1} at 25°C).

1.7 PRESSURE DEPENDENCE OF RATE CONSTANTS

It is possible to measure the rate of a reaction at various applied pressures and determine the variation of k with P. Such measurements have become increasingly widespread in recent years for a variety of inorganic reactions, and the interpretation of the pressure dependence adds a further tool to the arsenal of parameters available for mechanistic interpretations.

The variation of K with P is given by the van't Hoff equation:

$$\left(\frac{\delta \ln K}{\delta P} \right)_T = -\frac{\Delta V^o}{RT} \qquad (1.60)$$

where ΔV^o is the difference in the partial molar volume between the products and reactants. Since $K = k_1/k_{-1}$, it is logical to express the variation of k_1 with P by

$$\left(\frac{\delta \ln k_1}{\delta P}\right)_T = -\frac{\Delta V_1^*}{RT} \tag{1.61}$$

where ΔV_1^* is defined as the volume of activation for the forward step and is equal to the partial molar volume of the transition state minus the partial molar volume of the reactant(s). If ΔV_1^* is independent of pressure, then integration of Eq. (1.61) at constant temperature over the limits $P = 0$ to P and $k_1 = (k_1)_0$ to k_1 gives

$$\ln k_1 = \ln(k_1)_0 - \frac{\Delta V_1^* P}{RT} \tag{1.62}$$

so that a plot of $\ln k_1$ versus P should be linear with a slope of $- \Delta V_1^*/RT$. Since these studies usually cover pressures up to several thousand atmospheres, $\ln(k_1)_0$ is taken as the value at ambient pressure. If the plot is not linear, it is assumed that the reactant and/or transition state may be compressible and their volumes as a function of pressure can be described by

$$\Delta V_1^* = (\Delta V_1^*)_0 - \Delta \beta^* P \tag{1.63}$$

where $\Delta \beta^*$ represents the compressibility of the system. Then, substitution of Eq. (1.63) into Eq. (1.61) and integration over the same limits yields

$$\ln k_1 = \ln(k_1)_0 - \frac{(\Delta V_1^*)_0 P}{RT} + \Delta \beta^* \left(\frac{P^2}{2RT}\right) \tag{1.64}$$

The pressure dependence for the isomerization of $(NH_3)_5Co(ONO)^{2+}$ is shown in Figure 1.4. For the k_{OH} path, the authors[10] have ascribed the slight curvature of the plot to a compressibility effect with $\Delta \beta^* = 5$ cm^3 kbar^{-1} mol^{-1}. The difference between the upper straight line and the dashed curve shows the extent of the deviation from linearity.

1.8 IONIC STRENGTH DEPENDENCE OF RATE CONSTANTS

For reactions of ions in solution, the variation of the activity coefficients with reagent concentrations is sometimes ignored or, more commonly, assumed to be held constant by carrying out the reaction in the presence of some "inert electrolyte," which is at a much higher concentration than that of the reactants.

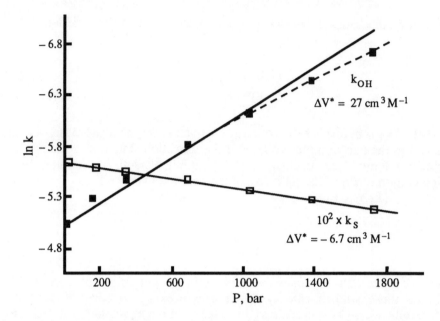

Figure 1.4. The pressure dependence of the linkage isomerization of $(NH_3)_5Co(ONO)^{2+}$.

The kinetic effect of ionic strength can be illustrated for a bimolecular reaction of the following type:

$$A + B \; \xrightleftharpoons{K^*} \; \{ A\text{----}B \}^* \; \longrightarrow \; \text{Products} \qquad (1.65)$$

for which transition-state theory predicts that the rate is given by

$$\text{Rate} = \frac{kT}{h} K^* \left(\frac{\gamma_A \, \gamma_B}{\gamma^*} \right) [A][B] = k[A][B] \qquad (1.66)$$

where γ_A, γ_B, and γ^* are the activity coefficients of the reactants and transition state, respectively. At infinite dilution (zero ionic strength), the activity coefficients are 1 and the rate constant is defined as k_0; therefore,

$$k = \left(\frac{\gamma_A \, \gamma_B}{\gamma^*} \right) k_0 \qquad (1.67)$$

The simplest relationship between the activity coefficients (γ_i) and μ is given by the Debye–Hückel limiting law, which applies for $\mu \leq 0.01$ M:

$$\log \gamma_i = -A z_i^2 \sqrt{\mu} \tag{1.68}$$

where A is a constant for a given solvent (A = 0.509 $M^{-1/2}$ for water at 25°C) and z_i is the charge of the ion. Using this limiting law, and realizing that $z^* = z_A + z_B$, it follows from Eq. (1.67) and (1.68) that

$$\log k = \log k_0 + 2 A z_A z_B \sqrt{\mu} \tag{1.69}$$

and a plot of log k versus $\sqrt{\mu}$ should be linear with a slope of $2 A z_A z_B$.

Many kinetic studies are done at ionic strengths beyond the range of applicability of the Debye–Hückel limiting law. The law was extended by Debye and Hückel to take into account the finite size of the ions to give Eq. (1.70), which is applicable for $\mu < 0.1$ M:

$$\log k = \log k_0 + \frac{2 A z_A z_B \sqrt{\mu}}{1 + \alpha B \sqrt{\mu}} \tag{1.70}$$

where α is the average effective diameter of the ions and B is a constant depending on the solvent properties (B = 0.328 $Å^{-1}$ $M^{-1/2}$ for water at 25°C). Values of α for various ions in water have been tabulated by Klotz.[11] Empirical equations have been developed for higher ionic strengths, such as the Davies equation,[12] which is applicable for $\mu \leq 0.5$ M.

The ionic strength dependence of k is essentially a property of the rate law. Therefore, the ionic strength dependence seldom affords new mechanistic information unless the complete rate law cannot be determined. These equations more often are used to "correct" rate constants from one ionic strength to another for the purpose of rate constant comparison. Ionic strength effects have been used to estimate the charge at the active site in large biomolecules, but the theory is substantially changed[13] because the size of the biomolecule violates basic assumptions of Debye–Hückel theory.

1.9 DIFFUSION-CONTROLLED RATE CONSTANTS

The upper limit on a rate constant for a reaction is imposed by the rate at which the reactants can diffuse together. This limit can be of significance when a particular mechanism would require a rate constant beyond the diffusion-controlled limit; then, the mechanism can be eliminated as a reasonable possibility. In addition, certain classes of reactions are known to proceed at or near diffusion-controlled rates, and this information can be useful in constructing and analyzing mechanistic models.

If two reactants A and B with radii r_A and r_B diffuse together and react at an interaction distance ($r_A + r_B$), then theories developed from Brownian motion predict that the diffusion controlled second-order rate constant is given by

$$k_{diff} = \frac{4\pi N(D_A + D_B)(r_A + r_B)}{1000}\left(\frac{U}{\exp(U)-1}\right) \tag{1.71}$$

where

$$U = \frac{z_A z_B e^2}{4\pi\varepsilon_0\varepsilon(r_A + r_B)kT}$$

D_A and D_B are the diffusion coefficients for A and B, respectively, z_A and z_B are their charges, N is Avogadro's number, k is Boltzmann's constant (1.38×10^{-23} J K^{-1}), e is the electron charge (1.6×10^{-19} C), ε is the dielectric constant of the solvent, and $4\pi\varepsilon_0 = 1.11 \times 10^{-10}$. If one or both of the species are neutral, then $U = 0$ and the right-hand term in brackets in Eq. (1.71) is equal to 1.

The diffusion coefficients can be approximated from the Stokes–Einstein equation, $D = kT/6\pi\eta r$, where $k = 1.38 \times 10^{-16}$ erg K^{-1} and η is the solvent viscosity , so that

$$k_{diff} = \frac{2RT(r_A + r_B)^2}{3000\eta r_A r_B}\left(\frac{U}{\exp(U)-1}\right) \tag{1.72}$$

with r in centimeters, η in poise, and $R = 8.31 \times 10^7$ erg mol^{-1} K^{-1}. This equation shows that k_{diff} will be relatively independent of the size of the reactants as long as $r_A \approx r_B$ and its magnitude will depend inversely on the solvent viscosity. The temperature dependence of k_{diff} will be governed largely by that of the solvent viscosity, so that apparent activation energies for diffusion-controlled processes are found to be in the 1 to 3 kcal mol^{-1} range for common solvents.

One can estimate k_{diff} from Eq. (1.72) without knowing the diffusion coefficients. Some values for various reactant sizes, assuming $r_A = r_B$ and charge products in water ($\eta = 0.00894$ poise, $\varepsilon = 78.3$), are given in Table 1.1. It is apparent from these data that diffusion-controlled rate constants in water can be expected to be in the range of 10^9 to 10^{10} M^{-1} s^{-1}.

For a unimolecular dissociation reaction, such as A—B forming A + B, the rate is controlled by the diffusion of the products out of the solvent cage. Theory predicts that the limiting dissociation rate constant is given by Eq. (1.73) which can be further simplified using the Stokes–Einstein equation.

$$k_{diff} = \frac{3(D_A + D_B)}{(r_A + r_B)^2}\left(\frac{U}{1-\exp(-U)}\right) \tag{1.73}$$

The predicted unimolecular dissociation rate constants (s^{-1}) are of the same magnitude as the bimolecular constants in Table 1.1.

Table 1.1. Estimated Diffusion Controlled Rate Constants (25°C) in Water

$r_A = 5$ (Å) $z_A \times z_B$	$r_B = 2.0$ (Å)	$r_B = 5.0$ (Å) k_{diff} (M^{-1} s^{-1})	$r_B = 8.0$ (Å)
−2	9.15 x 10^9	7.44 x 10^9	7.85 x 10^9
−1	9.10 x 10^9	7.42 x 10^9	7.83 x 10^9
0	9.05 x 10^9	7.39 x 10^9	7.81 x 10^9
+1	9.01 x 10^9	7.36 x 10^9	7.78 x 10^9
+2	8.96 x 10^9	7.34 x 10^9	7.76 x 10^9

The most important general class of reactions that have diffusion-controlled rates in water are protonation of a base by H_3O^+ and deprotonation of an acid by OH^-, as shown in the following reactions, with rate constants at 25°C in M^{-1} s^{-1}:

$$B: + H_3O^+ \xrightleftharpoons{k \approx 4 \times 10^{10}} B:H^+ + H_2O \qquad (1.74)$$

$$A:H + OH^- \xrightleftharpoons{k \approx 1 \times 10^{10}} A:^- + H_2O \qquad (1.75)$$

These results stem from the pioneering work of Eigen and co-workers.[14] The reverse rate constants for these reactions can be calculated from the equilibrium constants that are known for a wide range of such acids and bases. It is important to note that the reverse rate constants may not be extremely large. For example, trimethylamine is a strong base with a protonation rate constant of 6 x 10^{10} M^{-1} s^{-1}, but the K_a of the trimethylammonium ion is 1.6 x 10^{-10} M, so that the reverse reaction (deprotonation by water) has k = (6 x 10^{10})(1.6 x 10^{-10}) ≈ 100 s^{-1}.

The main exceptions to the preceding generalizations are so-called carbon acids, such as nitromethane or acetylacetone, for which the rate constants are usually much smaller and dependent on the nature of the acid. Such reactions are thought to be slower because of the bonding rearrangements required at carbon as the conjugate base is formed. The same constraint may apply to the deprotonation of organometallic hydrides.

References
1. Moore, J. W.; Pearson, R. G. *Kinetics and Mechanism*, 3rd ed.; Wiley-Interscience: New York, 1980.
2. Laidler, K. J. *Chemical Kinetics*, 3rd ed.; Harper & Row: New York, 1987.

3. Wilkins, R. G. *Adv. Inorg. Bioinorg. Mech.* **1983**, *2*, 139.
4. Caldin, E. F. *Fast Reactions in Solution*; Wiley & Sons: New York, 1964.
5. *Comprehensive Chemical Kinetics*; Bamford, C. H.; Tipper, C. F. H., Eds.; Elsevier: Amsterdam, 1969; Vol. 1, 2.
6. Mangelsdorf, P. C. *J. Appl. Phys.* **1959**, *30*, 443.
7. Espenson, J. E. *Chemical Kinetics and Reaction Mechanisms*; McGraw-Hill: New York, 1981.
8. Rodiguin, N. M.; Rodiguina, E. N. *Consecutive Chemical Reactions*, English Ed., translated by R. F. Schneider; Van Nostrand: Princeton, N.J., 1969.
9. Capellos, C.; Bielski, B. H. *Kinetic Systems*; McGraw-Hill: New York, 1972.
10. Jackson, W. G.; Lawrance, G. A.; Lay, P. A.; Sargeson, A. M. *Inorg. Chem.* **1980**, *19*, 904.
11. Klotz, I. *Chemical Thermodynamics*; Prentice-Hall: New York, 1950; pp. 330–331.
12. Davies, C. W. *Prog. React. Kin.* **1961**, *1*, 129.
13. Rosenberg, R. C.; Wherland, S.; Holwerda, R. A.; Gray, H. B. *J. Am. Chem. Soc.* **1976**, *98*, 6364.
14. Eigen, M. *Angew. Chem., Int. Ed.* **1964**, *3*, 1.

2

Rate Law and Mechanism

Once the experimental rate law has been established, the next step is to formulate a mechanism that is consistent with the rate law. The rate law will not uniquely define the mechanism but will limit the possibilities. The proposed mechanism will lead to predictions of trends in reactivity and other types of experiments that can be done to test the proposal.

Except for the simplest cases, the development of the rate law from the mechanism can be a messy exercise. The following sections describe some of the assumptions and tricks that can be used. Further discussions can be found in standard textbooks on kinetics.[1-3]

2.1 STEADY-STATE APPROXIMATION

A mechanism often invokes an unstable intermediate of some defined structure, and a general mechanism might take the form of

$$A \underset{k_2}{\overset{k_1}{\rightleftharpoons}} \{B\} \underset{k_4}{\overset{k_3}{\rightleftharpoons}} C \tag{2.1}$$

where B is an unstable and therefore reactive intermediate. The steady-state approximation assumes that this intermediate will disappear as quickly as it is formed:

$$\text{Rate of Appearance of B} = \text{Rate of Disappearance of B} \tag{2.2}$$

so that

$$k_1 [A] + k_4 [C] = k_2 [B] + k_3 [B] \tag{2.3}$$

With Eq. (2.3), one can solve for [B] in terms of the reactant and product concentrations ([A] and [C]) to give

$$[B] = \frac{k_1 [A] + k_4 [C]}{k_2 + k_3} \tag{2.4}$$

The total concentration of reagents can be defined as [T] and will remain constant. Since B is a reactive intermediate, its concentration will always be small relative to [A] + [C], so that

$$[T] = [A] + [B] + [C] = [A] + [C] \qquad (2.5)$$

Substitution for [C] in Eq. (2.4) allows one to obtain [B]:

$$[B] = \frac{k_1 [A] + k_4 [T] - k_4 [A]}{k_2 + k_3} \qquad (2.6)$$

The rate of disappearance of A is given as follows (note that the mechanism has specified that all the steps are first-order or pseudo-first-order):

$$-\frac{d[A]}{dt} = k_1 [A] - k_2 [B] = \frac{(k_2 k_4 + k_1 k_3)[A] - k_2 k_4 [T]}{k_2 + k_3} \qquad (2.7)$$

where the steady-state expression for [B] has been used to eliminate [B] from the differential equation; this gives a form that is integratable because [A] is the only concentration variable. However, instead of integrating at this point, it is useful to introduce the equilibrium (final) concentrations $[A]_e$ and $[C]_e$ through

$$[T] = [A]_e + [C]_e \qquad (2.8)$$

and

$$K_e = \frac{[C]_e}{[A]_e} = \frac{k_1 k_3}{k_2 k_4} \qquad (2.9)$$

Substitution for $[C]_e$ from Eq. (2.9) into Eq. (2.8) and rearranging gives

$$k_2 k_4 [T] = (k_1 k_3 + k_2 k_4)[A]_e \qquad (2.10)$$

Substitution for [T] from Eq. (2.10) into Eq. (2.7) yields

$$-\frac{d[A]}{dt} = \frac{(k_2 k_4 + k_1 k_3)}{(k_2 + k_3)} ([A] - [A]_e) \qquad (2.11)$$

which is the mathematical equivalent of the first-order rate law and can be integrated directly to obtain

$$-\ln([A]_e - [A]) + \ln([A]_e - [A]_0) = k_{exp} t \qquad (2.12)$$

where

$$k_{exp} = \frac{(k_2 k_4 + k_1 k_3)}{(k_2 + k_3)} \tag{2.13}$$

Note that the right-hand side of Eq. (2.13) is the same as the coefficient for [A] in Eq. (2.7) and that it was not really necessary to go through the equilibrium conditions in order to find the expression for k_{exp}. It is always true that once one has an integratable equation with only one concentration variable in first-order form, the coefficient of the concentration variable will be the expression for k_{exp}.

A limiting form of Eq. (2.13) that is often encountered assumes that $k_2 \gg k_3$. Then, the k_{exp} is given by

$$k_{exp} = \left(\frac{k_1}{k_2}\right) k_3 + k_4 = K_{12} k_3 + k_4 \tag{2.14}$$

The steady-state approximation can be applied to systems with any number of reactive intermediates. King and Altman[4] have presented a general development for k_{exp} in steady-state systems that is very useful for complex reaction networks.

2.2 RAPID EQUILIBRIUM ASSUMPTION

The rapid equilibrium treatment assumes that the reactants are part of a rapidly attained equilibrium that is always maintained during the course of the reaction, as shown by

$$A \underset{}{\overset{K_{12}}{\rightleftharpoons}} B \xrightarrow{k_3} C \tag{2.15}$$

where B is not a reactive intermediate but might be, for example, the conjugate base of A, a structural isomer of A, or an ion pair. Species B may be present at significant stoichiometric concentrations. The total concentration of the reactant can be defined as [R], where

$$[R] = [A] + [B] \tag{2.16}$$

Since [T] = [A] + [B] + [C], it follows that

$$[R] = [T] - [C] \tag{2.17}$$

Normally, one will know [R] but not [A] or [B], unless K_{12} is known.

The rate of formation of C is easily written down as

$$\frac{d[C]}{dt} = k_3 [B] \tag{2.18}$$

and the problem is to express [B] in terms of [C], in order to obtain an equation that can be integrated. A useful trick can be used to get an expression for the concentration of one of the partners in the equilibrium ([B]) in terms of the total concentration of the species involved in the equilibrium ([R]). Since $K_{12} = [B]/[A]$, then

$$\frac{1}{K_{12}} + 1 = \frac{[A]+[B]}{[B]} = \frac{[R]}{[B]} \tag{2.19}$$

Rearrangement and substitution for [R] from Eq. (2.17) into Eq. (2.19) gives

$$[B] = \frac{[R]\,K_{12}}{(K_{12}+1)} = \frac{([T]-[C])\,K_{12}}{(K_{12}+1)} \tag{2.20}$$

Substitution for [B] into Eq. (2.18) yields an equation with only [C] as the concentration variable:

$$\frac{d[C]}{dt} = \frac{k_3\,K_{12}}{(K_{12}+1)}([T]-[C]) \tag{2.21}$$

so that

$$k_{exp} = \frac{k_3\,K_{12}}{(K_{12}+1)} \tag{2.22}$$

The expression for k_{exp} may be compared to that derived from the steady-state assumption under the condition that $k_2 \gg k_3$. The k_4 is missing in the present example because we have assumed an irreversible model, but otherwise the steady-state and equilibrium models are the same if $K_{12} \ll 1$ (in which case the concentration of B is small).

The preceding discussion can leave the incorrect impression that B is like a particularly stable intermediate on the reaction pathway from A to C. A somewhat different perspective is gained if one views B as the starting material and A as some unreactive form of B. This situation produces the same rate law as Eq. (2.21). *The important general lesson is that all rapid equilibria involving the reactant(s) will enter into the rate law, even if the species involved are not on the net reaction pathway.*

2.3 PRINCIPLE OF DETAILED BALANCING

The principle of detailed balancing states that when a system is at equilibrium, the rate in the forward direction equals the rate in the reverse direction for each individual step in the process as well as for the overall reaction.

This can be of use in simple systems because it makes it possible to express one of the rate constants in terms of the others and the overall equilibrium constant. For the reaction

$$A + B \underset{k_r}{\overset{k_f}{\rightleftharpoons}} C + D \qquad (2.23)$$

$K_e = k_f/k_r$, so that $k_r = k_f/K_e$. For cyclic systems, a less obvious consequence is that the product of the rate constants going in one direction around the cycle must equal the product of rate constants in the other direction.

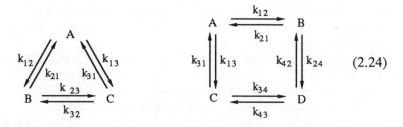

$$(2.24)$$

For the three species system, $k_{12}k_{23}k_{31} = k_{13}k_{32}k_{21}$, so that one needs to know only five of the six rate constants in order to define the system. Similarly, for the four-species system, one obtains $k_{13}k_{34}k_{42}k_{21} = k_{12}k_{24}k_{43}k_{31}$.

2.4 PRINCIPLE OF MICROSCOPIC REVERSIBILITY

The mechanism of a reverse reaction must be the same as the mechanism of the forward reaction under the same conditions. This results because *the least energetic pathway in one direction must be the least energetic pathway in the other direction.* The intermediates and transition state must be the same in either direction. One consequence of this is that a catalyst for a forward reaction will be a catalyst for the reverse reaction.

The proper application of the principles of microscopic reversibility and detailed balancing can be helpful in mechanistic assessments, as shown by recent discussion of the CO exchange in $Mn(CO)_5X$ systems. Johnson and co-workers[5] initially claimed that all the CO ligands were being exchanged at a similar rate and proposed the mechanism in Scheme 2.1.

Scheme 2.1

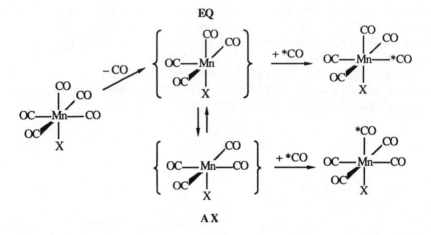

Brown[6] pointed out that this mechanism violates the principle of microscopic reversibility because, if dissociation of a cis CO is kinetically more favorable, then addition of a CO to the cis position must also be more favorable.

Subsequent work[7] using infrared detection indicates that the exchange of the cis CO is faster. Jackson[8] has suggested that the more recent analysis transgresses the principle of detailed balancing, but this criticism arises from an incorrect extension[9] of Brown's arguments by Espenson.[10] The detailed analysis by Jackson, allowing for initial dissociation of both cis- and trans-CO ligands, shows that the ratio of cis to trans products is independent of time if the intermediates are in rapid equilibrium, but the ratio varies with time otherwise, unless the two dissociation rates happen to be equal.

References
1. Moore, J. W.; Pearson, R. G. *Kinetics and Mechanism*, 3rd ed.; Wiley-Interscience: New York, 1980.
2. Laidler, K. J. *Chemical Kinetics*, 3rd ed.; Harper & Row: New York, 1987.
3. Espenson, J. E. *Chemical Kinetics and Reaction Mechanisms*; McGraw-Hill: New York, 1981.
4. King, E. L.; Altman, C. *J. Phys. Chem.* **1956**, *60*, 1375.
5. Johnson, B. F. G.; Lewis, J.; Miller, J. R.; Robinson, B. H.; Robinson, P. W.; Wojcicki, A. *J. Chem. Soc. A* **1968**, 522.
6. Brown, T. L. *Inorg. Chem.* **1968**, *7*, 2873.
7. Atwood, J. T.; Brown, T. L. *J. Am. Chem. Soc.* **1975**, *97*, 3380.
8. Jackson, W. G. *Inorg. Chem.* **1987**, *26*, 3004,
9. Brown, T. L. *Inorg. Chem.* **1989**, *28*, 3229.
10. Espenson, J. H. *Chemical Kinetics and Reaction Mechanisms*; McGraw-Hill: New York, 1981; pp. 128–131.

3

Ligand Substitution Reactions

In ligand substitution reactions one or more ligands around a metal ion are replaced by other ligands. In many ways, all inorganic reactions can be classified as either substitution or oxidation–reduction reactions, so that substitution reactions represent a major type of inorganic process. Some examples of substitution reactions follow:

$$Fe(OH_2)_6^{3+} + SCN^- \rightleftharpoons (H_2O)_5Fe-SCN^{2+} + H_2O$$

$$Ni(OH_2)_6^{2+} + en \rightleftharpoons \left(\begin{array}{c} H_2O \quad H_2 \\ N^-CH_2 \\ | \quad | \\ H_2O-Ni-N^-CH_2 \\ H_2O \quad | \quad H_2 \\ OH_2 \end{array} \right)^{2+} + 2 H_2O$$

(3.1)

$$Cr(CO)_6 + P(C_6H_5)_3 \longrightarrow \begin{array}{c} CO \quad CO \\ | \; / \\ OC-Cr-P(C_6H_5)_3 \\ / \quad | \\ OC \quad CO \end{array} + CO$$

3.1 OPERATIONAL APPROACH TO CLASSIFICATION OF SUBSTITUTION MECHANISMS

The operational approach was first expounded in 1965 in a monograph by Langford and Gray.[1] It is an attempt to classify reaction mechanisms in relation to the type of information that kinetic studies of various types can provide. It delineates what can be said about the mechanism on the basis of the observations from certain types of experiments. The mechanism is classified by two properties, its stoichiometric character and its intimate character.

Stoichiometric Mechanism

The stoichiometric mechanism can be determined from the kinetic behavior of one system. The classifications are as follows:

1. *Dissociative* (**D**): an intermediate of lower coordination number than the reactant can be identified.
2. *Associative* (**A**): an intermediate of larger coordination number than the reactant can be identified.
3. *Interchange* (**I**): no detectable intermediate can be found.

Intimate Mechanism

The intimate mechanism can be determined from a series of experiments in which the nature of the reactants is changed in a systematic way. The classifications are as follows:

1. *Dissociative activation* (**d**): the reaction rate is more sensitive to changes in the leaving group.
2. *Associative activation* (**a**): the reaction rate is more sensitive to changes in the entering group.

This terminology has largely replaced the S_N1, S_N2, and so on, type of nomenclature that is still used in physical organic chemistry. These terminologies are compared and further explained as follows:

Dissociative [**D** $\equiv$ S_N1 (limiting)]: there is definite evidence of an intermediate of reduced coordination number. The bond between the metal and the leaving group has been completely broken in the transition state without any bond making to the entering group.

Dissociative interchange ($I_d \equiv S_N1$): there is no definite evidence of an intermediate. In the transition state, there is a large degree of bond breaking to the leaving group and a small amount of bond making to the entering group. The rate is more sensitive to the nature of the leaving group.

Associative interchange ($I_a \equiv S_N2$): there is no definite evidence of an intermediate. In the transition state, there is some bond breaking to the leaving group but much more bond making to the entering group.

Associative [**A** $\equiv$ S_N2 (limiting)]: there is definite evidence of an intermediate of increased coordination number. In the transition state, the bond to the entering group is largely made while the bond to the leaving group is essentially unbroken.

The general goal of a kinetic and mechanistic study of a substitution reaction is to classify the reaction as **D**, I_d, I_a, or **A**.

3.2 OPERATIONAL TESTS FOR THE STOICHIOMETRIC MECHANISM

According to the original definitions, it should be possible to establish the stoichiometric mechanism on the basis of a study of one system. In practice, this has been expanded to include studies with one metal ion complex system.

3.2.a Dissociative Mechanism Rate Law

The dissociative mechanism can be described by the following sequence of reactions:

$$R\!-\!X \underset{k_2}{\overset{k_1}{\rightleftharpoons}} \{R\} + X \xrightarrow[Y]{k_3} R\!-\!Y \qquad (3.2)$$

where X and Y are the leaving and entering group, respectively, and $\{R\}$ is the intermediate of reduced coordination number.

The k_{exp} for the limiting dissociative mechanism can be derived from the previous solution of the system $A \rightleftharpoons B \rightleftharpoons C$ with a steady state for B (Eq. (2.13) with $k_4 = 0$). Then, if $[X]$ and $[Y] \gg [RX]$, k_{exp} is given by

$$k_{exp} = \frac{k_1 k_3 [Y]}{k_2 [X] + k_3 [Y]} = \frac{k_1 [Y]}{\dfrac{k_2}{k_3} [X] + [Y]} \qquad (3.3)$$

If this rate law is to provide a successful test of the **D** mechanism, it is necessary for the conditions to be such that $k_2[X] \approx k_3[Y]$. Then, for example, if $[X]$ is held constant and $[Y]$ is varied in a series of experiments, k_{exp} should change with $[Y]$, as shown in Figure 3.1. This type of variation is often referred to as "saturation" behavior, and k_{exp} approaches a limiting value of k_1 when $k_3[Y] \gg k_2[X]$. It is possible to rearrange Eq. (3.3) to give

$$\left(k_{exp} \right)^{-1} = \frac{k_2}{k_1 k_3} \left(\frac{[X]}{[Y]} \right) + \frac{1}{k_1} \qquad (3.4)$$

Therefore, a plot of $(k_{exp})^{-1}$ versus $[X]/[Y]$ should be linear and k_1 and k_2/k_3 can be determined.

If several different entering groups (Y) are studied they should all yield the same value of k_1 as a further condition of a **D** mechanism.

3.2.b Ion Pair or Preassociation Problem

The success of the preceding test of a **D** mechanism depends on the assumption that there are no other reaction sequences that produce the same rate law. Unfortunately, this is *not* true.

Many metal ion complexes are positively charged and many common entering groups are anions. Such oppositely charged species can form association complexes commonly called ion pairs. The phenomenon of preassociation is not limited to ions and may be significant for polar species

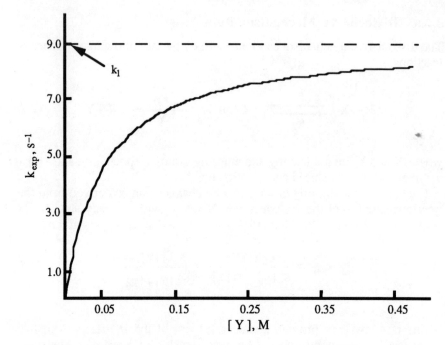

Figure 3.1. Predicted variation of k_{exp} with [Y] for a **D** mechanism with $k_1 = 9 \ s^{-1}$ and $k_2[X]/k_3 = 0.1$ M.

in nonpolar solvents due to dipole–dipole interactions and hydrogen bonding.

The general process can be described by the following sequence of reactions:

$$R—X + Y \ \underset{}{\overset{K_i}{\rightleftharpoons}} \ (R—X{\cdot}Y) \ \xrightarrow{k_3} \ R—Y + X \qquad (3.5)$$

where $(R—X{\cdot}Y)$ is the ion pair or preassociation complex formed in a fast pre-equilibrium with an equilibrium constant K_i. The rate law for this type of system has been developed previously in Eq. (2.22) and, if $[Y] \gg [RX]$, the pseudo-first-order rate constant is given by

$$k_{exp} = \frac{k_3 \, K_i \, [Y]}{K_i \, [Y] + 1} \qquad (3.6)$$

This equation predicts the same type of variation of k_{exp} with [Y] as that from the **D** mechanism if [X] is constant in Eq. (3.3). The latter is often the case because X is the solvent.

Table 3.1. Calculated Ion Pair Formation Constants $(a = 5 \times 10^{-8}$ cm, $T = 298$ K)

Ionic Strength $z_1 z_2$	H_2O ($\varepsilon = 78.5$)			CH_3OH ($\varepsilon = 32$)		CH_2Cl_2 ($\varepsilon = 9.1$)
	0.01	0.10	1.0	0.01	0.10	0.01
-1	1.08	0.81	0.54	5.1	2.2	1.3×10^3
-2	3.67	2.07	0.93	83	15	5.2×10^6
-3	12.5	5.3	1.6	1400	104	2.1×10^{10}
-4	42.7	13.6	2.7			
-5	146	34.7	4.7			
-6	497	88.9	8.1			

For ion pairs, Eigen[2] and Fuoss[3] developed an equation to estimate K_i based on extended Debye–Hückel theory and a hard-sphere model for the ions. It is given by

$$K_i = \frac{4 \pi N a^3}{3000} \exp\left(-\frac{U}{kT}\right) \qquad (3.7)$$

where

$$U = \frac{z_1 z_2 e^2}{\varepsilon}\left(\frac{1}{a(1+\kappa a)}\right) \quad \text{and} \quad k = \sqrt{\frac{8 \pi N e^2 \mu}{1000 \varepsilon kT}}$$

and N is Avogadro's number (6.022×10^{23}), a is the distance of closest approach of the ions (cm), e is the electron charge (4.803×10^{-10} esu), k is Boltzmann's constant (1.38×10^{-16} erg K^{-1}), ε is the bulk solvent dielectric constant, μ is the ionic strength, and z_1, z_2 are the charges on the ions. Some calculated values of K_i are given in Table 3.1. Experience indicates that these calculated values are reasonable approximations when compared to the few experimental values. The main point to note is that the value of $K_i[Y]$ can easily be of the same magnitude as 1 for typical charge types and for reasonable concentrations of Y.

3.2.c Competition Studies for the Intermediate

These studies attempt to test the prediction of a D mechanism that a particular metal center should produce the same intermediate independent of the leaving group. For example, one might study the solvolysis in water of $(NH_3)_5CoCl^{2+}$ and $(NH_3)_5CoONO_2^{2+}$, where Cl$^-$ and NO$_3^-$ are the leaving groups, in the presence of some added nucleophile Y. The object is to get the same product distribution if a common intermediate $\{(NH_3)_5Co^{3+}\}$ is formed. The confidence in the conclusions depends on studying a significant range of leaving groups.

The principles of the method are described by the following sequence:

$$
R\!-\!X \longrightarrow \{R\} + X
\begin{cases}
\xrightarrow[k_w]{H_2O} R\!-\!OH_2 \\
\\
\xrightarrow[k_y]{Y} R\!-\!Y
\end{cases}
\tag{3.8}
$$

The product ratio $[RY]/[ROH_2]$ can be calculated as follows:

$$
\frac{d[RY]}{dt} = k_y[R][Y] \quad \text{and} \quad \frac{d[ROH_2]}{dt} = k_w[R][H_2O]
$$

$$
\frac{d[RY]}{d[ROH_2]} = \frac{k_y}{k_w}\frac{[Y]}{[H_2O]} = \frac{k_y[Y]}{k_w'}
\tag{3.9}
$$

Integration and rearrangement yields

$$
\frac{[RY]}{[ROH_2][Y]} = \frac{k_y}{k_w'}
\tag{3.10}
$$

If a **D** mechanism is operative, the ratio on the left should be a constant for different concentrations of Y and for different leaving groups.

A similar analysis can be applied if the product complex has different structural isomers or stereoisomers. Then, the isomers should be produced in a proportion independent of the leaving group.

3.2.d Constant Thermodynamic Properties for the Intermediate

In this approach, it is hoped to show that the thermodynamic parameters of the intermediate are constant and independent of the leaving group, and thereby establish the independent nature of the intermediate.

The following development is in terms of enthalpy, but the same can be done for free energy, entropy, partial molar volume, and so on. The reaction energetics are defined by Figure 3.2. From this diagram it should be apparent that

$$
\Delta H^* - \Delta H_{stab} = \Delta H_f^o(R) + \Delta H_f^o(X) - \Delta H_f^o(RX)
\tag{3.11}
$$

For the overall reaction $RX + Y \rightarrow RY + X$, the enthalpy change is

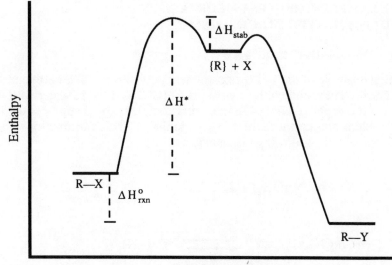

Figure 3.2. Reaction coordinate diagram for a **D** mechanism.

$$\Delta H^o_{rxn} = \Delta H^o_f(RY) + \Delta H^o_f(X) - \Delta H^o_f(RX) - \Delta H^o_f(Y) \qquad (3.12)$$

Therefore, the last two equations can be combined to eliminate $\Delta H^o_f(RX)$ and $\Delta H^o_f(X)$ and give

$$\Delta H^o_f(R) = \Delta H^* - \Delta H_{stab} - \Delta H^o_{rxn} + \Delta H^o_f(RY) - \Delta H^o_f(Y) \qquad (3.13)$$

If a series of leaving groups are examined using the same Y (e.g., the solvent), then

$$\Delta H^o_f(R) = \Delta H^* - \Delta H_{stab} - \Delta H^o_{rxn} + Constant \qquad (3.14)$$

This equation is not truly independent of X because of ΔH_{stab}, but this term is assumed to be small, so that

$$\Delta H^* - \Delta H^o_{rxn} \approx \Delta H^o_f(R) - Constant \qquad (3.15)$$

For a **D** mechanism, $\Delta H^* - \Delta H^o_{rxn}$ is expected to be constant for a particular entering group.

3.3 EXAMPLES OF TESTS FOR A DISSOCIATIVE MECHANISM

3.3.a   Dissociative Rate Law

The first example of the full **D** rate law was published by Wilmarth and co-workers.[4,5] They studied the anation of $Co(CN)_5OH_2{}^{2-}$ in water with the idea that the negative charge on the metal complex would suppress the ion pair problem and might favor a **D** mechanism. Their observations were consistent with the following mechanism:

$$Co(CN)_5OH_2{}^{2-} \underset{k_2}{\overset{k_1}{\rightleftharpoons}} \{Co(CN)_5{}^{2-}\} + H_2O$$

$$\{Co(CN)_5{}^{2-}\} + Y^- \underset{k_4}{\overset{k_3}{\rightleftharpoons}} Co(CN)_5Y^{3-}$$

(3.16)

The predicted pseudo-first-order rate constant for this mechanism can be obtained by analogy to Eq. (2.13), to give

$$k_{exp} = \frac{k_1 k_3 [Y] + k_2 [H_2O] k_4}{k_2 [H_2O] + k_3 [Y]} = \frac{k_1 k_3 [Y] + k_2' k_4}{k_2' + k_3 [Y]}$$

(3.17)

The value of k_4 was determined independently by studying the rate of aquation of $Co(CN)_5Y^{3-}$ with $[Y] \approx 0$, in which case $k_{exp} = k_4$. If k_4 is subtracted from both sides of the k_{exp} expression and the reciprocal is taken, then one obtains Eq. (3.18) which predicts that a plot of $(k_{exp} - k_4)^{-1}$ versus $[Y]^{-1}$ should be linear, allowing one to calculate k_1 and k_2/k_3:

$$(k_{exp} - k_4)^{-1} = (k_1 - k_4)^{-1} + \frac{k_2'}{k_3}(k_1 - k_4)^{-1} [Y]^{-1}$$

(3.18)

Wilmarth and co-workers found that their data satisfied this rate law and yielded reasonably constant values of k_1 for a range of Y such as Br^-, NH_3, I^-, SCN^-, and N_3^-. Note that, if Y is H_2O, then $k_{exp} = k_1$, so that a study of the water exchange rate would provide a further test.

   Unfortunately, all of the preceding results have been thrown into serious doubt by recent work. Burnett and Gilfillian[6] and then Haim[7] found that the rate law with $Y = N_3^-$ has a simple first-order dependence on $[N_3^-]$. Haim's observations indicate that the early work may be in error because of the presence of $(NC)_5CoO_2Co(CN)_5{}^{6-}$, which has been avoided in the recent studies through modified preparative procedures. The original observations with regard to SCN^- have been confirmed by later work.[8]

Table 3.2. Values of k_3/k_2 for Systems with a **D** Mechanism Rate Law

Entering Group	$Rh(Cl)_5OH_2{}^{2-}$ in Water	Entering Group	$(L)Cr(TPP)Cl$ in Toluene
H_2O	1.0	Pyridine	1.0
I^-	0.018	$P(Ph)_3$	0.0017
Br^-	0.016	$P(C_2H_4CN)_3$	0.0085
Cl^-	0.021	$P(OPr)_3$	0.075
SCN^-	0.079	N-Methylimidazole	1.7
$NO_2{}^-$	0.10		
$N_3{}^-$	0.14		

The water exchange rate on $Co(CN)_5OH_2{}^{2-}$ has been measured recently by Swaddle and co-workers[9] who find $k_{exch} = 5.8 \times 10^{-4}$ s^{-1} at 25°C, with $\Delta H^* = 90.2$ kJ mol^{-1} and $\Delta S^* = -4$ J mol^{-1} K^{-1}. This predicts that at 40°C, $k_{exch} = 3.5 \times 10^{-3}$ s^{-1} whereas the results of the substitution studies give $k_1 = 2 \times 10^{-3}$ s^{-1} (Haim) or 6×10^{-4} s^{-1} (Burnett).[10]

The current status of this system is that it is probably using a dissociative interchange mechanism (I_d) and that there is some preassociation of the metal complex and the entering group despite their unfavorable charge product.

Still, there are some systems for which the rate law indicates a dissociative mechanism. Some examples are $Rh(Cl)_5OH_2{}^{2-}$,[11] $en_2Co(SO_3)(OH_2)^+$,[12] bis(dimethylglyoxime) cobalt(III) complexes,[13] and tetraphenylporphine chromium(III) chloride (LCr(TPP)Cl).[14] The values of k_3/k_2 in these systems are of interest as they represent the relative reactivity of the entering to the leaving group. Such information could be relevant to general nucleophilicity scales, which will be discussed later. Some values are given in Table 3.2.

3.3.b Competition Studies for a Dissociative Intermediate

The main limitation for these studies is that the products must be stable enough that their amounts can be accurately determined. The favorite systems for these studies have been cobalt(III) amine complexes, because of their stability and the extensive documentation of their properties.

The early work was done on the hydroxide ion catalyzed hydrolysis of cobalt(III) amines, for which there was evidence that the reaction proceeds by a dissociative conjugate base mechanism (S_N1CB in earlier terminology), as shown in Scheme 3.1.

An analysis of early results on the hydrolysis of cis and trans isomers of $(en)_2Co(Y)(X)$ complexes by Sargeson and Jordan[15] indicated that if Y is the same, then the percentage of cis and trans isomers in the product

Scheme 3.1

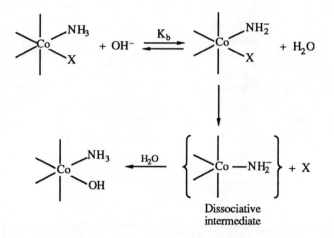

Dissociative
intermediate

(en)$_2$Co(Y)(OH) is fairly constant. Further work by Buckingham et al.[16] on stereoisomers of *cis*-(en)$_2$Co(NH$_3$)(X) is summarized in Table 3.3. The percentage of cis and trans products is quite constant, but the percentage retention is significantly greater with neutral leaving groups. This can be rationalized if the anionic leaving groups are retained longer within the immediate solvation sheath of the "intermediate" and tend to inhibit entry from the position they have vacated, thereby giving less retention. In the same study, using azide ion as a competing ligand, it was found that neutral leaving groups give about 5 percent more (en)$_2$Co(NH$_3$)(N$_3$)$^{2+}$ than anionic leaving groups. The preceding rationale can also be used to explain this observation.

The current status of this and related work has been summarized recently by Jackson et al.[17] and the earlier observations have been revised and expanded by Buckingham and co-workers.[18] It now appears that an intermediate does form, but that it is very reactive and scavenges its immediate coordination sphere rather than sensing the stoichiometric amounts of various species in the bulk solution. The intermediate may not

Table 3.3. Product Distribution for *cis*-(en)$_2$Co(NH$_3$)(X) + OH$^-$

Leaving Group	% trans	% cis	% Retention	% Racemate
Cl$^-$	22	78	48	30
NO$_3^-$	23	77	47	30
Br$^-$	22	78	44	34
(H$_3$C)$_2$SO	23	77	52	25
(H$_3$CO)$_3$PO	23	77	54	23

be truly independent of its source nor of the "inert" ionic medium because it reacts while the leaving group is still in the vicinity, and the products essentially reflect the ionic atmosphere around the reactant. However, the leaving group does not dramatically change the reactivity pattern of the intermediate.

Basolo and Pearson[19] suggested that the $\{(NH_3)_4CoNH_2{}^{2+}\}$ intermediate is stabilized by back π bonding from the $NH_2{}^-$ ligand, as shown in the following structure:

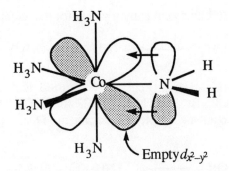

The competition results for the hydrolysis in acidic aqueous solution (aquation) show a greater sensitivity to the leaving group than those for base-catalyzed hydrolysis. The studies have also involved reactions designed to rapidly produce an intermediate of reduced coordination number, so-called induced aquations. The following are examples of typical induced reactions:

$$(NH_3)_5Co\text{—}Cl^{2+} + Hg^{2+} \longrightarrow \{(NH_3)_5Co^{3+}\} \xrightarrow[Z]{H_2O}$$
$$+ HgCl^+$$

$$(3.19)$$

$$(NH_3)_5Co\text{—}N_3{}^{2+} + NO^+ \longrightarrow \{(NH_3)_5Co^{3+}\} \xrightarrow[Z]{H_2O}$$
$$+ N_2 + N_2O$$

The intermediate here is different than that for the conjugate base mechanism, in that an $NH_2{}^-$ ligand is replaced by NH_3. It appears that the $\{(NH_3)_5Co^{3+}\}$ intermediate is more reactive than its conjugate base so that $\{(NH_3)_5Co^{3+}\}$ shows a greater dependence of the competition ratios on the leaving group. The induced aquation of $(NH_3)_5CoN_3{}^{2+} + NO^+$ in the presence of NCS^- yields 12 percent $(NH_3)_5Co(NCS)^{2+}$, whereas spontaneous aquation with trimethyl phosphate as the leaving group yields 4.6 percent $(NH_3)_5Co(NCS)^{2+}$. However, the distribution of linkage isomers in the product is nearly the same, with about 60 percent $(NH_3)_5Co\text{—}NCS^{2+}$ and 40 percent $(NH_3)_5Co\text{—}SCN^{2+}$. The status of $\{(NH_3)_5Co^{3+}\}$ is still a subject of controversy and is discussed in some detail in the recent paper by Jackson and Dutton[20] concerning the product distribution for the reaction of the azido complex with NO^+.

3.3.c   Constant Thermodynamic Parameters

As shown previously, for a **D** mechanism one can expect that

$$\Delta H^* - \Delta H^o_{rxn} = \Delta H^o_f(R) - \Delta H^o_f(RY) + \Delta H^o_f(Y) \tag{3.20}$$

House and Powell[21] analyzed enthalpy data for the aquation reaction

$$(NH_3)_5Co—X + OH_2 \longrightarrow (NH_3)_5Co—OH_2^{3+} + X \tag{3.21}$$

and found that $\Delta H^* - \Delta H^o_{rxn}$ (kcal mol^{-1}) varied with X, from 22.2 for SO_4^{2-} to ~25 for Cl^-, Br^-, and NO_3^-, to 27 for H_2O. This variation was taken as evidence against a simple **D** mechanism. On the other hand, for the reaction

$$(NH_3)_5Co—X + OH^- \longrightarrow (NH_3)_5Co—OH^{2+} + X \tag{3.22}$$

the same enthalpy difference is 32.4 ± 0.5 kcal mol^{-1} for the same range of ligands. This is quite constant and independent of the leaving group, therefore consistent with a **D** mechanism.

Activation volumes (ΔV^*) have been analyzed for the the same reactions[22,23] with similar conclusions. However, the analysis is more complex than was originally anticipated because of solvent electrostriction effects.

3.4   OPERATIONAL TEST OF AN ASSOCIATIVE MECHANISM

3.4.a   Associative Mechanism Rate Law

If the mechanism is represented by the reaction

$$R—X + Y \underset{k_2}{\overset{k_1}{\rightleftharpoons}} \left\{ R \begin{matrix} Y \\ X \end{matrix} \right\} \overset{k_3}{\longrightarrow} R—Y + X \tag{3.23}$$

and a steady state is assumed for the intermediate, and if $[Y] \gg [RX]$, then

$$k_{exp} = \frac{k_1 k_3 [Y]}{k_2 + k_3} = Constant [Y] \tag{3.24}$$

The rate should always be first order in [Y], and the rate law contains no information that uniquely defines an A mechanism.

A rather unlikely but feasible possibility is that the intermediate is formed in a rapid pre-equilibrium, as shown in the following:

$$R\!-\!X + Y \xrightleftharpoons{K_{12}} \left\{ R\!\!\underset{X}{\overset{Y}{\diagdown}} \right\} \xrightarrow{k_3} R\!-\!Y + X \qquad (3.25)$$

where, if [Y] >> [RX], then

$$k_{exp} = \frac{k_3 K_{12} [Y]}{K_{12} [Y] + 1} \qquad (3.26)$$

This expression has the same form as the one for the dissociative and ion pair rate laws. It might be distinguished from the latter by comparing the values of K_{12} to those expected for K_i. In addition, the spectral properties of the intermediate are likely to be much different from those of the reactant, whereas an ion pair is not much different because no bonds have been made or broken in the ion pair. The value of k_3 should depend on the nature of the entering group and thus could be distinguished from its mathematical equivalent k_1 in the dissociative rate law.

3.4.b   Examples of Associative Rate Laws

Examples of the full Associative rate law are rare because the "intermediate" must be quite stable if $K_{12}[Y] \geq 1$. Coordinatively unsaturated systems are most likely to satisfy this condition.

One apparent example was reported by Cattalini et al.[24] for the following reaction:

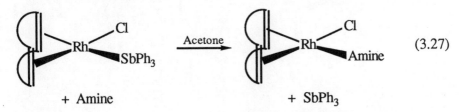

+ Amine   + SbPh$_3$

where the diene is cyclooctadiene. The rate is first-order in [Rh complex] and independent of [amine], the rate constant varies with the nature of the amine, and there is a rapid spectral change on mixing the reactants. If $K_{12}[amine] >> 1$, then $k_{exp} = k_3$ and should depend on the amine. The values of k_{exp}, 1.58×10^{-2} s^{-1} for 3-cyanopyridine to 4.57×10^{-2} s^{-1} for n-butylamine (in acetone at 25°C), do not show a large variation.

Another apparent example appears in the work of Toma and Malin[25] on the reaction

$$(H_3N)_5Ru-OH_2{}^{2+} + \quad N\overset{}{\underset{}{\bigcirc}}N^+-CH_3$$

(3.28)

$$\left((NH_3)_5Ru-N\overset{}{\underset{}{\bigcirc}}N^+-CH_3\right)^{3+} + H_2O$$

The dependence of k_{exp} on the concentration of methylpyrazinium ion is consistent with the Associative rate law. The authors argue that this is not due to ion pairing because of the like charges of the reactants. They suggest that the intermediate is a charge-transfer complex due to donation of t_{2g} electrons on Ru(II) to the empty π^* orbitals on the entering group. However, it is debatable whether this should be considered as an intermediate of expanded coordination number.

Species of expanded coordination number have been isolated and structurally characterized by Maresca et al.[26] as products of the reaction of Zeise's salt with bis hydrazones, as shown in Eq. (3.29). These five-coordinate products slowly lose ethylene in a first-order process.

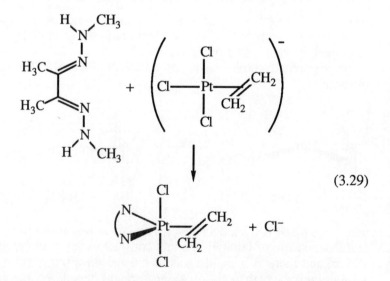

(3.29)

Tobe and co-workers[27] have provided evidence that reaction (3.30) proceeds through a stable intermediate, which may be the five-coordinate species shown or one of its structural isomers.

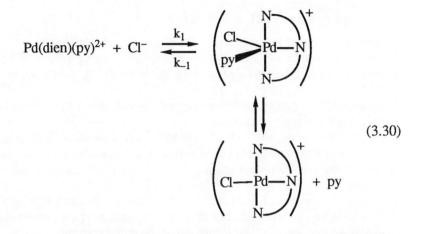

$$\text{Pd(dien)(py)}^{2+} + \text{Cl}^- \underset{k_{-1}}{\overset{k_1}{\rightleftharpoons}}$$

(3.30)

The intermediate forms reasonably quickly, with a rate that is first-order in [Cl$^-$] ($k_1 = 9.3 \times 10^{-2}$ M^{-1} s^{-1}, $k_{-1} = 2.5 \times 10^{-2}$ s^{-1} in 1 M NaClO$_4$ at 25°C), but too slowly to be considered as an ion pair. The ^{1}H nmr shows that the intermediate has not released pyridine nor undergone ring opening of the dien chelate.

3.5 OPERATIONAL TESTS FOR THE INTIMATE MECHANISM

These tests are concerned with the sensitivity of the reaction rate constant to the chemical nature of the entering and leaving groups for a general reaction, such as (3.31), in which there is no definite evidence for an intermediate:

$$\text{R---X} + \text{Y} \longrightarrow \text{R---Y} + \text{X} \qquad (3.31)$$

Associative activation requires more sensitivity to the nature of Y and dissociative activation requires more sensitivity to the nature of X.

These effects appear to be easy to test, but there is always a somewhat subjective decision in evaluating the degree of sensitivity to variations in X and Y. For example, in the associative case, since X is still present in the transition state, the rate constant must show some variation with X, but the variation with changes in Y must be greater. A further problem is that X or Y is often the solvent, and it cannot be changed without a major perturbation on the whole system. It is also necessary to ensure that the changes in X or Y have not been so trivial that the interaction with "R" in the transition state would not vary by much. For example, changing from Cl$^-$ to Br$^-$ would probably not produce much change in the rate constant for either type of activation. In order to avoid such possibilities, scales of nucleophilicity for various ligands are very useful. It is assumed that a better nucleophile will make a stronger bond to "R," so that one should choose entering or leaving groups of significantly different nucleophilicity in testing for the type of activation.

3.5.a   Inorganic Nucleophilicity Scales

Several variables are believed to generally affect nucleophilicity:

1. *Basicity towards H^+*: the commonly available pK_a values of ligands measure this and it seems to parallel the nucleophilicity toward many metal centers.

2. *Polarizability*: a more polarizable ligand should be a better electron donor and therefore a better nucleophile.

3. *Oxidizability*: a more easily oxidized ligand is more willing to give up electrons and therefore is expected to be a better nucleophile. This factor is measured by standard reduction potentials or polarographic half-wave potentials.

4. *Solvation energy*: a ligand that is more strongly solvated in a particular solvent will be a poorer nucleophile in that solvent because some solvation change must occur during the formation of the metal complex.

5. *Metal at reaction center*: this factor greatly limits the generality of nucleophilicity scales in inorganic chemistry when compared to those in organic chemistry.

3.5.a.i n_{Pt} Scale

The n_{Pt} scale is a kinetic scale based on the reaction

$$trans\text{-}(py)_2Pt(Cl)_2 + Y \xrightarrow{\text{Methanol}} trans\text{-}(py)_2Pt(Cl)(Y) + Cl^- \qquad (3.32)$$

with the definition that

$$n_{Pt} = \log\left(\frac{k_Y}{k_{\text{methanol}}}\right) \qquad (3.33)$$

Some typical values[28] of n_{Pt} are Cl^- (3.04); NH_3 (3.07); N_3^- (3.58), I^- (5.46); CN^- (7.14); PPh_3 (8.93). Clearly, if one is testing for entering group effects in Pt(II) chemistry, one should not choose Cl^- and NH_3 as test nucleophiles because their nucleophilicities are almost identical and one would not see much entering or leaving group effect. This scale works well for Pt(II) reactions, but is at best a qualitative indicator for other metals and not even that for the first-row transition-metal ions.

3.5.a.ii   Methyl–Mercury(II) Scale

The methyl–mercury(II) scale is based on the equilibrium constant for the reaction[29]

$$H_3C\text{—}Hg\text{—}Y^+ + H_2O \underset{}{\overset{K}{\rightleftharpoons}} H_3C\text{—}Hg\text{—}OH_2^+ + Y \qquad (3.34)$$

and nucleophilicity is taken as proportional to $-\log(K) = pK$. The pK values correlate well with the n_{Pt} scale and they have a similar range of applicability.

3.5.a.iii Edwards Scale

Edwards[30] proposed that a kinetic nucleophilicity scale could be based on a combination of the factors mentioned earlier for a particular metal center, using the equation

$$\log\left(\frac{k_Y}{k_{solvent}}\right) = \alpha(-E^o_{Y_2}) + \beta(pK_a + 1.74) \qquad (3.35)$$

where $E^o_{Y_2}$ is the reduction potential for $Y_2 + 2e^- \rightarrow 2Y^-$ in water. This scale is often mentioned but has limited applicability because of the lack of $E^o_{Y_2}$ values for all but the halogens and pseudohalogens.

3.5.a.iv Gutmann Donor Numbers

This scale is based on the enthalpy change for the reaction of Lewis bases with $SbCl_5$.[31] The donor number, $DN = -\Delta H^o_{rxn}$ (kcal mol^{-1}). The larger the DN the stronger the base and therefore the stronger the nucleophile. The scale has been expanded to include acceptor numbers, AN.

This scale is widely referenced in relation to thermodynamic arguments. It has been criticized because of the neglect of solvent effects and side reactions that contribute to ΔH^o_{rxn} and because a one-parameter scale can never be entirely adequate. Recent measurements[32] with BF_3 as the acid have provided some points of comparison and criticism for the original donor numbers. Donor numbers for some common solvents are given in Table 3.4.

3.5.a.v Hydrogen-Bonding Scales

The ability of a solvent to hydrogen bond with a particular donor can be monitored by various techniques, and this property has been suggested as a

Table 3.4. Donor Numbers for Common Solvents

Base	DN	Base	DN
Nitromethane	2.7	Ether	19.2
Acetonitrile	14.1	Methanol	20
Dioxane	14.8	Dimethylformamide	24
Acetone	17	Dimethylsulfoxide	29.8
Water	18	Pyridine	33

measure of solvent basicity. The two-parameter scale (ß and ζ)[33] of Kamlet and Taft covers a wide range of solvents and has been used and compared to other scales. The ζ parameter has been added as a covalency correction that depends on the type of acceptor atom. Taft and co-workers[34] have proposed hydrogen bond acidity parameters (α_2) that can be multiplied by the basicity parameter (β_2) to correlate a wide range of such acid–base equilibria.

A thermochemical scale of hydrogen bond basicity has been proposed based on the differences in the heats of solution ($\delta \Delta H°$) of pyrrole, *N*-methylpyrrole, benzene, and toluene.[35] The thermochemical scale correlates well with the ß parameter of Kamlet and Taft, with the exception of dioxane and especially triethylamine. A value for water, which is often absent in other scales, is provided by the thermochemical results.

Some results from these scales are given in Table 3.5. There has never been any claim that this type of scale measures basicity towards a metal, but one is often driven far afield in seeking such measures.

3.5.a.vi   Drago E and C Scale

The Drago E and C scale[36] is based on the enthalpy change for the interaction of a Lewis acid (A) and base (B). Each acid and base is characterized by two parameters, E_A and C_A for acids and E_B and C_B for

Table 3.5. Measures of Hydrogen Bond Basicity

Solvent (Base)	$\delta \Delta H°$ (kJ mol^{-1})	ß	$(\beta_{calc})^a$
Carbon tetrachloride	1.80		(−.09)
Benzene	5.67	0.10	(0.11)
Water	7.46		(0.20)
Nitromethane	8.82	0.25	(0.27)
Cyanobenzene	10.28	0.37	(0.35)
Acetonitrile	10.32	0.35	(0.35)
Methanol	12.93	0.43	(0.48)
Dioxane	13.15	0.37	(0.49)
Tetrahydrofuran	15.17	0.55	(0.60)
Pyridine	16.22	0.64	(0.65)
N,N-Dimethylformamide	16.61	0.69	(0.67)
Dimethylsulfoxide	17.99	0.74	(0.74)
Hexamethylphosphoramide	23.56	1.05	(1.03)
Triethylamine	24.11	0.71	(1.05)

^{a} Calculated from a linear correlation of $\delta \Delta H°$ and ß by Catalán, J.; Gómez, J.; Couto, A.; Laynez, J. *J. Am. Chem. Soc.* **1990,** *112,* 1678.

Table 3.6. E and C Values (kcal mol^{-1})$^{1/2}$ for Representative Acids and Bases

Acid	E_A	C_A	Base	E_B	C_B
I_2	1.0	1.0	NH_3	1.36	3.46
$SbCl_5$	7.38	5.13	$N(CH_3)_3$	0.81	11.5
$B(CH_3)_3$	6.14	1.70	$O=S(CH_3)_2$	1.34	2.85
$Al(CH_3)_3$	12.5	2.04	$O=C(CH_3)_2$	0.99	2.33
$Ga(CH_3)_3$	12.6	0.59	$O=P(CH_3)_3$	1.03	5.99
$Cu(hfac)_2$	3.46	1.36	$P(CH_3)_3$	0.84	6.55

bases, and the enthalpy change for the reaction A + B → (A : B) is given by

$$\Delta H^o_{rxn} = E_A E_B + C_A C_B \qquad (3.36)$$

The justification for this scale is that it is able to correlate a very large number and range of reaction enthalpies.

Drago has suggested that the C parameter is related to the covalent part of the interaction and that E is related to the ionic part. Therefore, a strong base that complexes with an acid through a largely ionic interaction will have a large E_B and probably a small C_B. These ideas are potentially useful in selecting appropriate nucleophiles for mechanistic tests, but have not been widely used. There has been a recent application to heterogeneous adsorption and catalysis.[37] Some values of E and C are given in Table 3.6.

3.5.a.vii Hard and Soft Acid–Base Theory

This terminology was first proposed by Pearson,[38] and the basic idea is related to the earlier separation of metal ions into (a) and (b) classes as suggested by Arhland et al.[39] Acids and bases were qualitatively classified by Pearson as "soft," "hard," or "borderline." Soft acids and bases were suspected of using covalent bonding in their interactions and hard species of using predominantly ionic forces. The rule of thumb was that *"hard acids interact most strongly with hard bases and soft acids interact most strongly with soft bases."* The main criticism of this scale is that it is purely qualitative. For mechanistic tests, it would be inappropriate to use a soft base with a hard acid because the interaction would be inherently weak, but there is no basis for selecting an appropriate range of nucleophiles to test for an I_a or I_d mechanism.

More recently, Pearson[40] has attempted to establish a quantitative scale of hardness and softness based on ionization potentials (I) and electron affinities (A). The absolute hardness is defined as $\eta = (I - A)/2$ and softness as $\sigma = 1/\eta$, and the absolute electronegativity uses Mulliken's definition $\chi = (I + A)/2$. For an interaction between a Lewis acid (1) and base (2), the strength of the interaction is assumed to be related to the fractional electron transfer, given by $\Delta N = (\chi_1 - \chi_2)/2(\eta_1 + \eta_2)$. These results are too recent to have been tested, except to note that the quantitative scale generally conforms to the ideas of practicing chemists. Pearson has compiled an extensive list of hardness values and a few examples are given in Table 3.7.

The following selection gives a general idea of the types of hard and soft acids and bases:

Hard Acids:	H^+, Li^+, Mg^{2+}, Cr^{3+}, Co^{3+}, Fe^{3+}
Soft Acids:	Cu^+, Ag^+, Pd^{2+}, Pt^{2+}, Hg^{2+}, Tl^{3+}
Borderline:	Mn^{2+}, Fe^{2+}, Zn^{2+}, Pb^{2+}

Hard Bases:	F^-, Cl^-, H_2O, NH_3, OH^-, $H_3CCO_2^-$
Soft Bases:	I^-, CO, $P(C_6H_5)_3$, C_2H_4, H_5C_2SH
Borderline:	N_3^-, C_5H_5N, NO_2^-, Br^-, SO_3^{2-}

3.5.a.viii   Summary

None of these scales has received universal acceptance by inorganic chemists and it may be that the heterogeneity of the field will defy anyone to establish a truly general scale. As yet, there seems to be nothing as widely applicable as the Hammet and Taft parameters in organic chemistry. Within certain

Table 3.7. Absolute Electronegativity (χ) and Hardness (η) Values (eV)

Acid	χ	η	Base	χ	η
Ca^{2+}	31.39	19.52	F^-	10.41	7.01
Cr^{2+}	23.73	7.23	Cl^-	8.31	4.70
Mn^{2+}	24.66	9.02	Br^-	7.60	4.24
Fe^{2+}	23.42	7.24	I^-	6.76	3.70
Co^{2+}	25.28	8.22	OH^-	7.50	5.67
Zn^{2+}	28.84	10.88	H_2O	3.1	9.5
Fe^{3+}	42.73	12.08	NH_3	2.6	8.2
Ru^{3+}	39.2	10.7	$N(CH_3)_3$	1.5	6.3
Os^{3+}	35.2	7.5	C_5H_5N	4.4	5.0
Co^{3+}	42.4	8.9	CH_3CN	4.7	7.5
Cr^{3+}	40.0	9.1	$(CH_3)_2O$	2.0	8.0
Pd^{2+}	26.18	6.75	$(CH_3)_3P$	2.8	5.9
Pt^{2+}	27.2	8.0	$(CH_3)_2NCHO$	3.4	5.8

areas and types of applications one finds one of these scales more often used than others, presumably because it has proved more successful in correlating information.

It has been recognized that a two-parameter scale is necessary, in general, to correlate solvent basicity.[41,42] The conditions under which a one-parameter correlation may appear to work have been discussed by Drago[43] and the correlations between various basicity scales have been analyzed by Maria et al.[44]

In mechanistic studies, it is common to fall back on qualitative information relating to the particular metal or nonmetal center. After working with particular types of compounds, a lore develops about what are good, not-so-good, and poor nucleophiles. Sometimes, it is recognized that this correlates with one of the basicity scales.

3.5.b Linear Free-Energy Relationships (LFER)

The relationship between the equilibrium constant and the rate constants for a simple reaction can be used for mechanistic purposes. Since $K_e = k_f / k_r$,

$$\log(k_f) = \log(K_e) + \log(k_r) \tag{3.37}$$

If a reaction has dissociative activation and one varies the nature of the leaving group X while keeping the entering group Y constant, then k_r should be constant and a plot of $\log(k_f)$ versus $\log(K_e)$ should be linear with a slope of +1. One can do the converse type of experiment and plot for associative activation.

This type of analysis has been applied to the aquation of $(NH_3)_5Co^{III}$—X complexes[45,46] with the conclusion that the mechanism is I_d. The LFER for $(NH_3)_5Cr^{III}$—X and $(H_2O)_5Cr^{III}$—X, indicates an I_a mechanism.[47]

Proton transfer reactions are an important class that generally satisfy a LFER. Although they are not directly related to ligand substitution processes, proton transfer steps are often involved and usually treated as rapidly maintained equilibria:

$$HB + H_2O \xrightleftharpoons[k_r]{k_f} B^- + H_3O^+ \tag{3.38}$$

$$HB + OH^- \xrightleftharpoons[k_r]{k_f} B^- + H_2O \tag{3.39}$$

For reactions such as (3.38), k_r is fairly constant at ~5 x 10^{10} M^{-1} s^{-1}, the diffusion-controlled limit as noted on p. 19. This is not true only when HB and B^- have significant structural and bonding differences. If the acid dissociation constant (K_a) is known, then $k_f \approx 5 \times 10^{10} \times K_a$. Note that, if

K_a is small (e.g., 10^{-10} M), then k_f can be rather modest and the fast equilibrium assumption may not be valid. The same analysis applies to reactions such as (3.39), where $k_f \approx 2 \times 10^{10}$ M^{-1} s^{-1}.

3.5.c   Reagent Charge Effects

It is often possible to change the net charge on a metal complex by changing the "nonreacting" ligands (e.g., $PtCl_4^{2-}$, $Pt(NH_3)(Cl)_3^-$, $Pt(NH_3)_2(Cl)_2$, etc.). The variation of substitution rate constant with charge appears to give a clear distinction between associative and dissociative activation. Increasing positive charge on the metal complex should favor bonding to the entering nucleophile and therefore increase the rate of an I_a process, whereas the opposite would be expected for an I_d process.

Unfortunately, when the charge is changed the ligands must change, and these types of studies can be difficult to interpret. For the Pt(II) examples given earlier, other evidence indicates an I_a mechanism, but there are only minor differences[48] in the aquation rates. It can be argued that the increasing positive charge increases the bonding to the entering group but has a similar effect on the leaving group, and the effects tend to cancel.

3.5.d   Solvent Effects

In organic systems, an increase of rate with increasing dielectric constant of the solvent is associated with the formation of an electrically polar transition state. The situtation is more complex with inorganic systems where the metal complex and the leaving and entering groups are often charged, and one must consider both desolvation of the reactants and solvation of the transition state. The dielectric constant of the solvent also has a substantial influence on the formation of ion pairs that may affect the apparent reactivity. A further complication is that the solvent may be a potential ligand and therefore part of the reacting system rather than just a reaction medium. As a result, the effect of solvent variation on the rate has not been a generally useful criterion for mechanism. More commonly, such observations are used to assess the solvent effect, when the mechanism is thought to be known, and to test this variation for various theoretical models.

In a system discussed later, Wax and Bergman[49] used a series of methyl-substituted tetrahydrofurans, of presumed constant dielectric constant, to test for the involvement of coordinated solvent in the "intermediate" formed after ligand dissociation. In a study more related to an organic chemistry type of application, Rerek and Basolo[50] attempted to use solvent variations to differentiate between the following rhodium intermediates. Clearly, the Rh$^\pm$ intermediate is more polar and should be favored by solvents with higher dielectric constants. The observations are given in Table 3.8. The authors favor the Rh$^\pm$ intermediate on the basis of a comparison of the rate constants in THF and methanol. But when the data are presented as in Table 3.8, it could be argued that the solvents in the left-hand column show an inverse

Table 3.8. Variation of the Rate Constant (25°C) with the Solvent Dielectric Constant for the Reaction of $(\eta^x\text{-}C_5H_4NO_2)Rh(CO)_2$ with PPh_3

Solvent	ε	$k\ (M^{-1}\ s^{-1})$	Solvent	ε	$k\ (M^{-1}\ s^{-1})$
Hexane	1.88	4.44	Dichloromethane	8.93	2.81
Cyclohexane	2.02	3.92	Methanol	32.7	10.4
Toluene	2.38	1.26	Acetonitrile	38.8	9.58
Tetrahydrofuran	7.58	0.963			

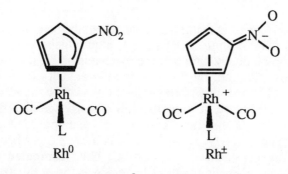

dependence of k on ε and favor Rh^0, whereas the solvents in the right-hand column might be going through a solvent-coordinated intermediate with methanol being the most strongly coordinating ligand.

3.5.e Steric Effects

Changes in the steric bulk of the nonreacting ligands appear to provide a clear distinction between I_d and I_a mechanisms. Increased steric bulk of the ligands should enhance an I_d mechanism by pushing the dissociating ligand away and relieving steric strain. On the other hand, it should make bonding with the entering group more difficult and inhibit an I_a mechanism. This is perhaps the most successful method of distinguishing these mechanisms, but there are some difficulties.

Parris and Wallace[51] studied the aquation of $M(NH_3)_5Cl^{2+}$ and $M(NH_2CH_3)_5Cl^{2+}$ with M = Co and Cr; the kinetic results of this and subsequent work are summarized in Table 3.9. The original interpretation of the rate constants was that the introduction of the $-CH_3$ group increases k for Co(III), consistent with an I_d mechanism, whereas the opposite effect is seen with Cr(III), as expected for an I_a mechanism for this metal center. This view has persisted until very recently and has caused many other types of information to be interpreted in a way consistent with this kinetic pattern for these two metal ions. Over the years, structural information has been accumulating and this has been analyzed recently by Lay.[52] The structures of the metal complexes are shown in Figure 3.3.

Table 3.9. Rate Constants (25°C) and Activation Parameters for the Aquation of Ammonia and Methylamine Complexes of Cobalt(III) and Chromium(III)

	$10^6 \times k$ (s^{-1})	ΔH^* $(kJ\ mol^{-1})$	ΔS^* $(J\ mol^{-1}\ K^{-1})$	ΔV^* $(cm^3\ mol^{-1})$
$Co(NH_3)_5Cl^{2+}$	1.72	93	−44	−2.1
$Co(NH_2CH_3)_5Cl^{2+}$	39.6	95	−10	~0
$Cr(NH_3)_5Cl^{2+}$	8.70	93	−29	−1.0
$Cr(NH_2CH_3)_5Cl^{2+}$	0.26	110	−2	~0

There are two structural features of note. First, the bond lengths to Co(III) are all shorter than those to Cr(III); this is a quite general feature and has been part of the rationalization that the mechanisms are different for these two metal ions. Second, the M—Cl bond is actually 0.03 Å shorter in $Cr(NH_2CH_3)_5Cl^{2+}$ than in $Cr(NH_3)_5Cl^{2+}$; this is opposite to what would have been predicted on steric grounds. Lay's interpretation of the structural and kinetic results is that both systems are reacting by a common I_d mechanism. The $Cr(NH_2CH_3)_5Cl^{2+}$ reacts more slowly because of the shorter, and presumably stronger, Cr—Cl bond compared to the NH_3 complex. Overall, there is really little evidence for steric strain in either

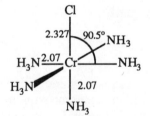
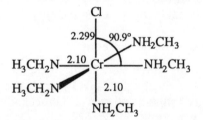
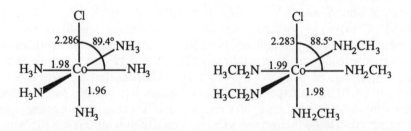

Figure 3.3. Structural parameters for cobalt(III) and chromium(III) amine complexes.

Table 3.10. Rate Constants for Replacement of the Chloro Ligand by Pyridine in $(Et_3P)_2Pt(R)(Cl)$

R	k (M^{-1} s^{-1})	
	trans (25°C)	cis (0°C)
⬡—Pt	1.2×10^{-4}	8×10^{-2}
⬡(CH$_3$)—Pt	1.7×10^{-5}	2×10^{-4}
H_3C—⬡(CH$_3$)(CH$_3$)—Pt	3.4×10^{-6}	1×10^{-6} (25°C)

system, and the nearly 90° bond angles confirm this impression. The $Co(NH_2CH_3)_5Cl^{2+}$ is more reactive because the ΔS^* is larger by 34 J mol^{-1} K^{-1} and not because the ΔH^* is smaller, as would have been expected if the Co—Cl bond were weaker in $Co(NH_2CH_3)_5Cl^{2+}$. Lay has suggested that the entropic difference is due to less effective solvation of the methylamine complex. The general conclusion is that if one is probing steric effects, one must be sure that the expected effects are present in the ground-state reactants.

In a somewhat different application of steric effects, Basolo et al.[53] studied reaction (3.40) with various R groups, with the results given in Table 3.10.

$$(Et_3P)_2Pt\,(R)\,(Cl) + py \xrightarrow{\text{EtOH}} (Et_3P)_2Pt\,(R)\,(py)^+ + Cl^- \quad (3.40)$$

The trend with both isomers is consistent with associative activation on these square planar Pt(II) complexes. The larger effect on the cis isomer can be rationalized by a trigonal bipyramidal intermediate, assuming that the entering group (Y) and leaving group (Cl) occupy equivalent positions in the trigonal plane in order to satisfy microscopic reversibility, as shown in Scheme 3.2. The R group in the axial position in the transition state for the cis isomer will cause more steric crowding than when it is in the equatorial position for the trans isomer.

Scheme 3.2

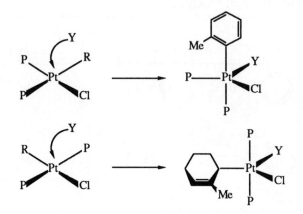

3.5.f Measures of Ligand Size: Cone Angles

The problem in interpreting the "steric" effects in the ammonia and methylamine complexes of cobalt(III) and chromium(III) might have been avoided if there were some method of estimating ligand size and therefore anticipating whether steric effects were really significant in a particular system.

The only systematic effort in this regard originates with the work of Tolman[54] on phosphine and phosphite ligands. Tolman defined a *cone angle* for a number of phosphines (PR_3) based on the following diagram, with the size of the R substituents based on van der Waals radii from CPK models. Some cone angle values are given in Table 3.11.

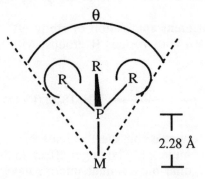

For unsymmetrical ligands P(R')(R")(R), the angle is calculated by

$$\theta = \frac{2}{3} \sum_{i=1}^{3} \frac{\theta_i}{2} \tag{3.41}$$

Table 3.11. Cone Angles and pK_a for Representative Phosphorus(III) Compounds[a]

Ligand	Cone Angle	pK_a
P(OMe)$_3$	107	2.6
PMe$_3$	118	8.65
P(Ph)Me$_2$	122	6.5
PCl$_3$	124	
P(OPh)$_3$	128	−2.0
PEt$_3$	132	8.69
P(Ph)$_2$Me	136	4.57
P(Ph)$_3$	145	2.73
P(CH$_2$Ph)$_3$	165	(6.0)
P(C$_6$H$_4$o-Me)$_3$	171	
P(C$_6$H$_4$o-Me)$_3$	194	3.08

[a] Rahman, Md. M.; Liu, H. Y.; Prock, A.; Giering, W. P. *Organometallics* **1987**, *6*, 650.

The original definition chose an M—P distance based on Ni(0) chemistry and applied the steric size of the ligands to correlate equilibrium constant data for the following reaction:

$$\text{Ni(L)}_4 \; \overset{K}{\rightleftharpoons} \; \text{Ni(L)}_3 + \text{L} \qquad (3.42)$$

Since the original application, these cone angles have been the basis for many attempts to correlate reactivity and steric effects. The original definitions have been criticized and examined by de Santo et al.,[55] and applied to more expanded correlations including electronic effects by Giering and co-workers.[56] There seems to have been little effort to extend the definition beyond phosphorus donor ligands, and the potential variation with M—P bond length is seldom taken into account.

The analysis by Giering emphasizes the fact that, below a certain cone angle threshold, steric effects should be minimal. They become significant when the sum of the cone angles for two adjacent ligands causes the ligands to come into contact.

3.5.g Volumes of Activation (ΔV^*)

This type of information has been discussed with regard to the **D** and **A** mechanisms, and the expectations for $\mathbf{I_d}$ and $\mathbf{I_a}$ are qualitatively similar. For the $\mathbf{I_d}$ mechanism, the prediction is that ΔV^* will be positive because the leaving group is being liberated into solution while there has been relatively

Table 3.12. Volumes and Entropies of Activation for Some Water Exchange Reactions

L_nM	ΔV^* ($cm^3 mol^{-1}$)	ΔS^* ($J mol^{-1} K^{-1}$)
$(NH_3)_5Co^{3+}$	+1.2	28
$(NH_3)_5Rh^{3+}$	−4.1	3
$(NH_3)_5Cr^{3+}$	−5.8	0
$(H_2O)_5Cr^{3+}$	−9.3	16
$(H_2O)_3Pt^{2+}$	−4.6	−9

little bonding to the entering group. On the other hand, for an I_a mechanism, the ΔV^* will be negative because the entering group has been captured from solution while the leaving group is still bonded to the metal center. Much of the literature implies this mechanistic differentiation is based on the sign (+ or −) of ΔV^*. It seems more correct to say that for a family of related reactions, those with a more positive ΔV^* are more dissociative and those with a more negative ΔV^* are more associative.

Values of ΔV^* for the following exchange reaction with M = Cr, Co, Rh[57] and M = Pt[58] are given in Table 3.12.

$$L_nM(OH_2) + H_2O^* \rightleftharpoons L_nM(^*OH_2) + H_2O \qquad (3.43)$$

The simplest explanation of this data is that the intimate mechanism involves increasing amounts of bond making to the entering H_2O as one goes from $(NH_3)_5Co^{3+}$ to $(H_2O)_5Cr^{3+}$, but it is *not* necessary that the former be I_d and the latter I_a. There is considerable other evidence that substitution on Pt(II) has an I_a mechanism, but the water exchange reaction may show less bond making than usual because Pt(II) is a soft acid and water is a hard base. It would be of great value to be able to predict the ΔV^* for a particular mechanism, but this has proved to be complicated because of solvent electrostriction effects. The solvent will constrict around the reactants as the charge density increases or will expand as the charge density decreases, and this factor is very difficult to anticipate in a quantitative way. Applications of ΔV^* are discussed further in the section "Other Theoretical Applications" later in this chapter and in a recent comprehensive review.[59]

3.5.h   Entropies of Activation (ΔS^*)

It is generally expected that an I_d mechanism will have a more positive ΔS^* than an I_a mechanism because the randomness increases in an I_d transition state and decreases in an I_a transition state. This criterion has been used cautiously and only for comparisons of similar reaction types. One reason for caution is the experimental error in ΔS^*, which is typically ± 8 to 12 J mol^{-1} K^{-1}. It was thought that ΔV^* would have an advantage over ΔS^* as a mechanistic criterion because of the better accuracy and our better intuitive understanding of volume changes. There is some evidence of a correlation between ΔS^* and ΔV^*,[60] which could make accurate ΔS^* values of greater mechanistic value. There is some indication of such a correlation in Table 3.12 for the systems with $L_n = (NH_3)_5$, but the $(H_2O)_n$ systems are different, possibly because of the usual scapegoat of solvation differences. The advantage of ΔS^* seems to be that the changes are much larger than with ΔV^* and this offsets the problem of experimental uncertainty.

3.6 SOME SPECIAL EFFECTS

3.6.a Internal Conjugate Base Mechanism

The internal conjugate base mechanism has been described previously in connection with competition studies for a **D** mechanism (Scheme 3.1). In its original form, it concerned the reaction of hydroxide ion with amine complexes of cobalt(III). It occupies a special place in inorganic mechanistic work because it was the subject of a long controversy during the 1960s between Ingold, Nyholm, and Tobe on one side and Basolo and Pearson on the other. This controversy stimulated a lot of kinetic work and also generated many useful spin-offs in the areas of synthesis, spectroscopy, and theory. Ingold and colleagues believed that the reaction was a simple bimolecular displacement (S_N2) of the leaving group by hydroxide ion, whereas Basolo and Pearson were proponents of the S_N1CB (DCB) mechanism originally proposed by Garrick.[61]

In the final analysis, the DCB mechanism has passed every test so far applied and is now the accepted mechanism for these reactions. The results leading to these conclusions are summarized in several articles.[62–64]

The DCB mechanism has received an amount of attention quite out of proportion to its general importance, since it is of less kinetic significance even for amine complexes of other 3+ metal ions. Its importance for cobalt(III) may be due to the small size of low-spin cobalt(III) and its ability to stabilize the dissociative intermediate through π bonding. However, it is generally observed that the conjugate bases of hydrated metal ions, shown in the following reaction, are much more reactive toward substitution:

$$M(OH_2)_n{}^{z+} \rightleftharpoons M(OH_2)_{n-1}(OH)^{(z-1)+} + H^+ \qquad (3.44)$$
$$\text{More active form}$$

3.6.b Trans Effect

It is well documented that the ligand in the position trans to the leaving group has a significant influence on the rate of substitution reactions. This has been most thoroughly documented in Pt(II) chemistry, where the ordering of trans labilization is

$$CN^-, CO > R_3P, H^-, (H_2N)_2CS > CH_3^- > NCS^- > I^- > Cl^- > OH^-$$

The sources of the trans effect can be classified into thermodynamic and kinetic factors. The thermodynamic factor refers to weakening of the bond to the leaving group in the ground state of the reactant. The kinetic factor refers to the stabilization of the transition state by the trans ligand.

3.6.b.i   Thermodynamic Trans Effect

Structural evidence for this effect was summarized by Appleton et al.[65] The bond length changes generally are not dramatic. For example, the Pt—Cl bond in *cis*-Pt(PR_3)_2(Cl)_2 is 0.08 Å longer than the Pt—Cl bond in *trans*-Pt(PR_3)_2(Cl)_2 although PR_3 is one of the stronger trans labilizing ligands. However H^- causes trans bond lengthening of 0.15 to 0.20 Å.[66] Indirect evidence for the thermodynamic trans effect was originally obtained by Chatt et al.[67] in T–Pt–NHR_2 systems from the N—H stretching frequencies. The C—O stretching frequency and nmr chemical shifts also have been used to probe for this effect.

It was recognized at an early stage that, in order to rationalize the order of the ligands in the trans effect series, it would be necessary to invoke both σ and π bonding effects. These effects work in the ground state through the metal orbitals shared by the leaving group (X) and the trans ligand (T), as shown in the following diagrams:

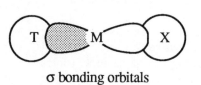

σ bonding orbitals

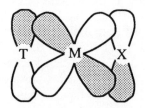

π bonding orbitals

The metal σ orbital is empty and both T and X donate electrons into it. If the T ligand is a stronger σ donor, then the σ bond to X will be weaker. The large trans effects of H_3C^- and H^- are directly attributed to this situation.

The metal π orbital is a filled d_{xy}, d_{xz}, or d_{yz} orbital and electrons are being donated from this orbital into empty orbitals on the ligands T and X. These empty orbitals might be π antibonding (π^*) on CO and CN^- or d orbitals on phosphorus-donor ligands, for example. This "back donation" strengthens the metal–ligand bond and, if T is a better π acceptor (π acid), then its bond to M will be strengthened at the expense of the M—X bond. This effect explains the positions of CO and PR_3 in the trans effect series, since these are typically good π acids.

In classic studies to separate the σ and π effects, Parshall[68] studied the
^{19}F nmr chemical shifts for a series of T ligands in the following types of
complexes:

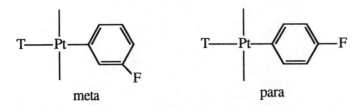

The ^{19}F shift in the meta isomer should sense only the σ effect, but the shift
in the para isomer will be sensitive to both the σ and π effects. Essentially
similar nmr methods have been applied to a number of different systems.
The results of these studies generally confirm the presence of both types of
effects, with magnitudes consistent with expectations for different T ligands.
The difficulty is that these observations do not translate easily into a direct
measure of the extent of bond weakening in the ground state caused by a
trans group; therefore the kinetic consequences are uncertain.

3.6.b.ii Kinetic Trans Effect

The kinetic trans effect is assumed to operate because of better bonding
between M and T in the transition state. The effect can be explained
pictorially, but there have been attempts at more quantitative quantum
mechanical calculations.[69,70] The contributions have been divided into σ
and π effects.

Most explanations have taken square planar Pt(II) complexes as models
because the effect is best documented for these systems and the mechanism
is taken to be I_a. The transition state is assumed to be a trigonal bipyramid
formed as follows:

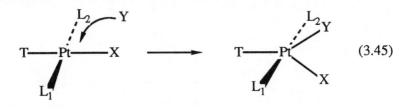

$$(3.45)$$

In the transition state, the T ligand is no longer sharing σ or π orbitals with
the X ligand, therefore the T ligand can improve its bonding to M if it is a
good σ donor or π acceptor. This will somewhat offset the loss in bonding
to X. In addition, in the transition state, the empty $d_{x^2-y^2}$ orbital on Pt(II)
becomes of π symmetry and can accept π electrons from T if the latter is
capable of π donation, as shown in the following diagram:

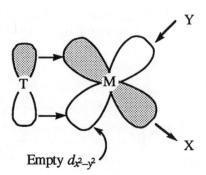

Empty $d_{x^2-y^2}$

3.6.c   Cis Effect

The effect of the cis ligand(s) on substitution rates has been noted especially in the chemistry of octahedral organometallic complexes. In general, the effect is less than the trans effect. The order of cis labilization for various ligands is

$$NO_3^- > CH_3CO_2^- > Cl^-, Br^- > Py > I^- > PR_3 > CO, H^-$$

This ordering is almost the exact opposite of that for the trans effect. The reaction mechanism also is different in that the systems used as examples show I_d characteristics. Evidence for a ground-state effect is very limited, as expected, because the cis ligands share only one orbital ($d_{x^2-y^2}$) of minor significance in their σ bonding. The effect is normally attributed to stabilization of the I_d transition state by better bonding with the cis ligand. Davy and Hall[71] have given a theoretical treatment of the cis effect for $Mn(CO)_5(L^-)$ and $Cr(CO)_5L$ systems. This approach indicates that the π-donor ability of the cis ligand is the most important feature, which is consistent with the ordering of ligands.

It should be remembered that for systems such as

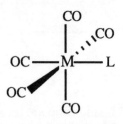

the observation is that the cis-CO ligands exchange or are replaced more easily than the trans one. This relative rate difference could be assigned, at least in part, to a trans *stabilizing* influence of the L group.

3.6.d   Reactions Without Metal–Ligand Bond Breaking

Some reactions of metal complexes that appear to involve substitution may actually occur without breaking a bond to the metal. The ^{18}O-labeling studies[72] for Eq. (3.46) show that the Co—O bond is retained, but the acid hydrolysis of the carbonato complex looks like simple aquation.

$$(NH_3)_5Co(^{18}OH)^{2+} + CO_2 \rightleftharpoons \left((NH_3)_5Co\overset{18}{-}O \overset{\diagdown}{\underset{O}{C}}-OH \right)^{2+}$$

$$\Updownarrow$$

$$(NH_3)_5Co(^{18}OCO_2)^+ + H^+$$

$$(3.46)$$

For chelated carbonate complexes, Posey and Taube[73] found that initial ring opening proceeds with Co—O bond breaking, and then CO_2 is lost by O—C bond breaking, as follows:

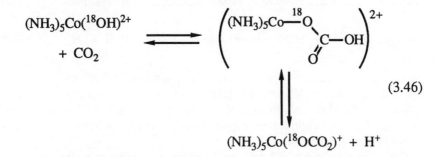

$$(3.47)$$

There is considerable interest in the reactions of such carbonato complexes because of their possible relevance to the action of carbonic anhydrase, and the area has been reviewed recently.[74]

Studies by Harris and co-workers[75] indicate that sulfur dioxide reacts in a fashion similar to CO_2, as shown by the example in Eq. (3.48). The immediate product is the O-bonded sulfito complex, which may undergo linkage isomerism to the S-bonded form or reduce cobalt(III) to cobalt(II).

$$(NH_3)_5Co(^{18}OH)^{2+} + SO_2 \longrightarrow \left((NH_3)_5Co\overset{18}{-}O \overset{\diagdown}{\underset{O}{S}}-OH \right)^{2+} \quad (3.48)$$

The aliphatic amine complexes react in this way, but the bis(phenanthroline) and bis(bipyridyl) complexes of cobalt(III) also show reactivity toward HSO_3^- and SO_3^{2-}.[76]

The reactions with nitrous acid follow a more complicated route, described by

$$2\ HNO_2 \rightleftharpoons N_2O_3 + H_2O \rightleftharpoons NO^+ + NO_2^- \qquad (3.49)$$

$$(NH_3)_5Co–OH^{2+} + NO^+ \longrightarrow (NH_3)_5Co–O–NO^{2+} + H^+ \qquad (3.50)$$

Murmann and Taube[77] found that the oxygen bonded to cobalt(III) in the reactant remains in the product, and it was long assumed that this oxygen remained bound to cobalt(III). However, recent work of Jackson et al.[78] indicates that the original cobalt-bound oxygen is scrambled between the two possible sites in the product.

The examples in Eqs. (3.46) and (3.48) can be viewed as intermolecular nucleophilic attack of coordinated OH^- on an electrophile. The same process can occur intramolecularly, as in the case of oxalate chelate ring opening studied by Andrade and Taube,[79] whose isotope tracer results are summarized by

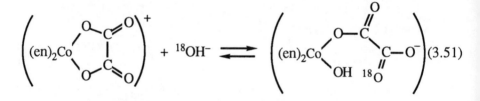

$$\qquad (3.51)$$

Microscopic reversibility requires that the ring-closing process must occur by O—C bond cleavage.

Similar intramolecular ring-closing, with retention of the Co—O bond, is observed with esters,[80] as shown in Eq. (3.52):

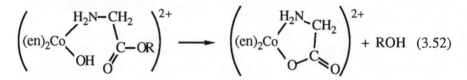

$$+\ ROH \qquad (3.52)$$

The amides[81] are more complex in that they can produce stable chelated amide or hydrolyzed amide products, as shown in Eq. (3.53), with the product distribution depending on the amide substituent.[82] In acidic solution, there is no general acid or base catalysis, and the rate-controlling step is attack of coordinated water on the carbonyl carbon followed by fast elimination of NH_2R. In alkaline solution, there is general base catalysis, which is interpreted as fast attack by coordinated OH^- followed by rate-controlling deprotonation of C—OH in the cyclic intermediate.

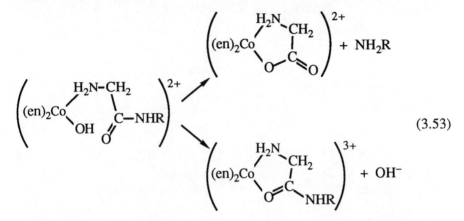

(3.53)

Bifunctional bases, such as HPO_4^{2-}, HCO_3^-, and $HAsO_4^{2-}$, are particularly good catalysts for amide hydrolysis. This has been attributed to the ability of such bases to simultaneously remove the C—OH proton and protonate the NHR leaving group, as shown in Eq. (3.54):

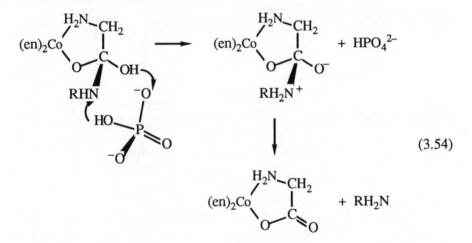

(3.54)

The chelate ring closing of glycine has been the subject of extensive isotope and kinetic studies,[83] and the results are summarized in Scheme 3.3 (k, s^{-1} at 25°C). The reactions proceed with retention of the Co—O bond, and it is noteworthy that coordinated water is an active nucleophile in this intramolecular process. The acidic pathways (k_1 and k_2) show general acid catalysis, which the authors interpret as due to rate-controlling elimination of water from the tetrahedral carbon in a cyclic intermediate.

Scheme 3.3

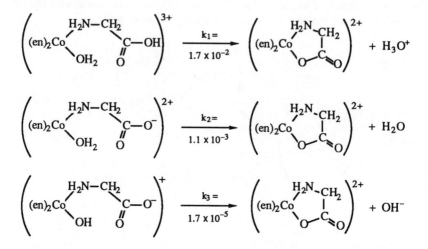

In other cases, coordination to a metal may activate the ligand to nucleophilic attack. The following equations give examples[84–87] of such processes that are generally much slower or not even observed with the free ligand:

$$(NH_3)_5Co{-}N{\equiv}C{-}CH_3^{3+} \xrightarrow{\ +OH^-\ } \left((NH_3)_5Co{-}NH{\diagdown}{\underset{O}{\overset{\|}{C}}}{\diagup}CH_3\right)^{2+} \quad (3.55)$$

$$(NH_3)_5Co{-}N{\equiv}C{-}N(CH_3)_2^{3+} \xrightarrow{\ +OH^-\ } \left((NH_3)_5Co{-}N{\overset{H}{\diagdown}}{\underset{O}{\overset{\|}{C}}}{-}N(CH_3)_2\right)^{2+} \quad (3.56)$$

$$(NH_3)_5Co{-}N{\equiv}C{-}CH_3^{3+} \xrightarrow{\ +N_3^-\ } \left((NH_3)_5Co{-}N{\diagup}{\underset{H_3C}{\diagdown}}{\overset{N{=}N}{\underset{C{=}N}{}}}\right)^{2+} \quad (3.57)$$

The following reaction, where tn is trimethylenediamine and RO^- is *p*-nitrophenolate, also gives chelate ring opening. The chelated phosphate ester hydrolyzes about 10^9 times faster than the free ester, and the acceleration is attributed to relief of ring strain in the five-coordinate phosphorus intermediate as well as to the inductive effect of the $(tn)_2Co^{3+}$.

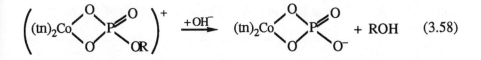

$$\left((tn)_2Co\begin{array}{c}O\\O\end{array}P\begin{array}{c}O\\OR\end{array}\right)^+ \xrightarrow{+OH^-} (tn)_2Co\begin{array}{c}O\\O\end{array}P\begin{array}{c}O\\O^-\end{array} + ROH \qquad (3.58)$$

It has been found[88] that chelated amino acid esters are activated so that they are useful for peptide synthesis, as shown in Eq. (3.59). This work has been recently reviewed.[89]

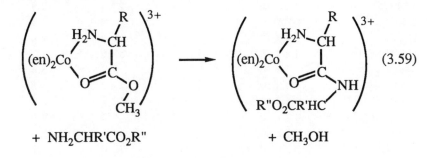

$$(3.59)$$

$$+ NH_2CHR'CO_2R'' \qquad\qquad + CH_3OH$$

3.7 VARIATION OF SUBSTITUTION RATES WITH METAL ION

The variation in rates of substitution with different metal ions has been an area of interest for many years. This type of information on the qualitative level is very useful for synthetic purposes and for studies on equilibrium properties. It provides an indication of the conditions required for a particular preparative procedure or of how long one must wait for a system to reach equilibrium. The wide range of rates for the transition-metal ions has been of considerable interest, both to rationalize the relative rates and to apply these rationalizations to the probable reaction mechanism.

3.7.a Water Solvent Exchange Rates

Since the pioneering work of Plane and Hunt[90] and of Swift and Connick,[91] one of the most widely studied reactions for transition-metal complexes has been that of water solvent exchange, using either $^{18}OH_2$ or $^{17}OH_2$:

$$L_nM(OH_2) + *OH_2 \rightleftharpoons L_nM(*OH_2) + H_2O \qquad (3.60)$$

The results of such studies are summarized in Table 3.13, and the field has recently been reviewed and discussed by Merbach.[92] Results of nmr studies prior to about 1970 should be viewed with caution, and most of the early work has been repeated with modern instrumentation and methods of analysis.

Table 3.13. Water Solvent Exchange Rate Constants (25°C) and Activation Parameters

$L_nM(OH_2)$	k (s^{-1})	ΔH^* (kJ mol^{-1})	ΔS^* (J mol^{-1} K^{-1})	ΔV^* (cm^3 mol^{-1})	Ref.
$Ti(OH_2)_6^{3+}$	1.8×10^5	43.4	1.2	-12.1	a
$V(OH_2)_6^{2+}$	8.7×10^1	61.8	-0.4	-4.1	b
$V(OH_2)_6^{3+}$	5.0×10^2	49.4	-27.8	-8.9	c
$Cr(OH_2)_6^{2+}$	$>10^8$				d
$Cr(OH_2)_6^{3+}$	2.4×10^{-6}	108.6	11.6	-9.6	e
$Cr(OH_2)_5OH^{2+}$	1.8×10^{-4}	111	55.6	2.7	e
$Mn(OH_2)_6^{2+}$	2.1×10^7	32.9	5.7	-5.4	f
$Fe(OH_2)_6^{2+}$	4.4×10^6	41.4	21.2	3.8	f
$Fe(OH_2)_6^{3+}$	1.6×10^2	63.9	12.1	-5.4	g
$Fe(OH_2)_5OH^{2+}$	1.2×10^5	42.4	5.3	7.0	g
$Co(OH_2)_6^{2+}$	3.2×10^6	46.9	37.2	6.1	f
$Ni(OH_2)_6^{2+}$	3.2×10^4	56.9	32.0	7.2	f
$Cu(OH_2)_6^{2+}$	$>10^7$				d
$Zn(OH_2)_6^{2+}$	$>10^7$				d
$Ru(OH_2)_6^{2+}$	1.8×10^{-2}	87.4	16.1	-0.4	l
$Ru(OH_2)_6^{3+}$	3.5×10^{-6}	89.8	-48.3	-8.3	l
$Ru(OH_2)_5OH^{2+}$	5.9×10^{-4}	95.8	14.9	0.9	l
$Pt(OH_2)_4^{2+}$	3.9×10^{-4}	89.7	-9	-4.6	k
$Pd(OH_2)_4^{2+}$	5.6×10^{-2}	49.5	-26	-2.2	k
$Cr(NH_3)_5OH_2^{3+}$	5.2×10^{-5}	97	0	-5.8	h
$Co(NH_3)_5OH_2^{3+}$	5.7×10^{-6}	111	28	1.2	i
$Rh(NH_3)_5OH_2^{3+}$	8.4×10^{-6}	103	3	-4.1	h
$Ir(NH_3)_5OH_2^{3+}$	6.1×10^{-8}	118	11	-3.2	j

[a] Hugi, A. D.; Helm, L.; Merbach, A. E. *Inorg. Chem.*, **1987**, *26*, 1763.

[b] Ducommun, Y.; Zbinden, D; Merbach, A. E. *Helv. Chim. Acta*, **1982**, *65*, 1385.

[c] Hugi, A. D.; Helm, L.; Merbach, A. E. *Helv. Chim. Acta*, **1985**, *68*, 508.

[d] Estimate from various kinetic studies.

[e] Xu, F.-C.; Krouse, H. R.; Swaddle, T. W. *Inorg. Chem.*, **1985**, *24*, 267.

[f] Ducommun, Y.; Newman, K. E.; Merbach, A. E. *Inorg. Chem.*, **1980**, *19*, 3696.

[g] Grant, M.; Jordan, R. B. *Inorg. Chem.*, **1981**, *20*, 55; Swaddle, T. W.; Merbach, A. E. *Inorg. Chem.*, **1981**, *20*, 4212.

[h] Swaddle, T. W.; Stranks, D. R. *J. Am. Chem. Soc.*, **1972**, *94*, 8357.

[i] Hunt, H. R.; Taube, H. *J. Am. Chem. Soc.*, **1958**, *80*, 2642.

[j] Tong, S. B.; Swaddle, T. W. *Inorg. Chem.*, **1974**, *13*, 1538.

[k] Helm, L.; Elding, L. I.; Merbach, A. E. *Inorg. Chem.*, **1985**, *24*, 1719.

[l] Rapaport, I.; Helm, L.; Merbach, A. E.; Bernhard, P.; Ludi, A. *Inorg. Chem.*, **1988**, *27*, 873.

A noteworthy aspect of the data in Table 3.13 is the wide range of rate constants: for the 2+ ions, k varies from 4×10^{-4} s^{-1} for Pd(II) to $>10^7$ s^{-1} for Cu(II) and Zn(II); for the hexaaqua 3+ ions, k varies from 3×10^{-6} s^{-1} for Cr(III) to 1×10^5 s^{-1} for Ti(III).

3.7.a.i Labile and Inert Classification of Taube

Taube[93] was the first to attempt an explanation of the substitution lability of these metal ions by classifying them qualitatively on the basis of their reactivity.

Labile metal ions react essentially on mixing of the metal ion and ligand solutions, that is, within a few seconds at most. *Inert metal ions* require at least a few minutes for their substitution reactions to be complete. This very operational classification was about all that was possible in 1952, but it provides a useful practical classification and this terminology has endured as a qualitative description of the reactivity of a metal ion.

Taube offered a theoretical explanation for the qualitative differences in reactivity. The original explanation was in terms of Pauling's valence bond theory, but the same arguments can be framed in terms of crystal field theory for octahedral complexes. The terminology is defined in Figure 3.4. *Labile metal ions* have either an empty low-energy t_{2g} orbital or at least one electron in a high-energy e_g orbital. The rationalization for this is that the empty t_{2g} orbital can be used by the entering group in an **A** or $\mathbf{I_a}$ transition state. Electrons in the higher-energy orbital set will favor a **D** or $\mathbf{I_d}$ mechanism because the ligand bonds in the ground state will be weaker. *Inert metal ions* must have at least one electron in each t_{2g} orbital and no electrons in the e_g orbitals.

These ideas are consistent with the inertness of the octahedral complexes of chromium(III) (t_{2g}^3), low-spin cobalt(III), iron(II) (t_{2g}^6) and iron(III) (t_{2g}^5), as well as the complexes of the second- and third-row transition metals with more than two d electrons, which generally are low-spin. They

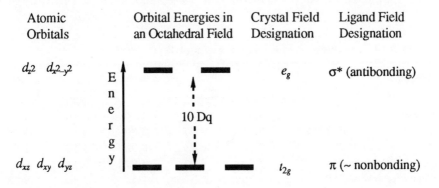

Atomic Orbitals   Orbital Energies in an Octahedral Field   Crystal Field Designation   Ligand Field Designation

d_{z^2} $d_{x^2-y^2}$ Energy   10 Dq e_g σ^* (antibonding)

d_{xz} d_{xy} d_{yz} t_{2g} π ($\sim$ nonbonding)

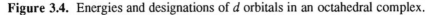

Figure 3.4. Energies and designations of d orbitals in an octahedral complex.

also provide an explanation for the fact that vanadium(III) (t_{2g}^2) is more labile than vanadium(II) (t_{2g}^3), whereas chromium(III) (t_{2g}^3) is inert and chromium(II) $(t_{2g}^3 e_g^1)$ is labile.

The predictions of this theory are qualitatively correct but it does not explain the wide range of reactivities, especially of the labile systems. For example, why is the water exchange rate for nickel(II) 10^2 times slower than that for cobalt(II) and $>10^4$ times slower than that for copper(II)?

3.7.a.ii   Crystal Field Theory Applications

This application of crystal field theory was put forward first in the textbook by Basolo and Pearson[94] in an attempt to explain the finer details of the reactivity differences between various metal ions.

The energies of the valence d orbitals were calculated for various ideal geometries of possible transition states for the substitution reactions. The calculations assume a pure crystal field model (no covalent bonding) and the same bond lengths and crystal field parameter (Dq) as in the ground state of the metal ion complex, with normal bond angles for the various geometries of the intermediates. The crystal field stabilization energy (CFSE) was calculated for the ground state and the intermediate, and the difference between these was defined as the *crystal field activation energy* (CFAE). The differences in reactivity were assigned to the difference in this electronic factor for various numbers of d electrons.

The energies of the d orbitals in units of Dq are given in Table 3.14. Based on these orbital energies, one can calculate the CFAE for each of the possible transition states and then predict the order of reactivity for the metal ions and the transition state that is most favorable (the one with the lower CFAE). Such calculations are shown for square pyramid and octahedral wedge intermediates in Table 3.15.

If one disregards differences in Dq for different metal ions, the preceding calculations predict that the order of reactivity for a square pyramid transition state should be $(d^4, d^9) > (d^2, d^7) > (d^1, d^6) > (d^0, d^5, d^{10}) > (d^3, d^8)$.

Table 3.14. Energies of d Orbitals in Units of Dq

Structure	$d_{x^2-y^2}$	d_{z^2}	d_{xy}	d_{xz}	d_{yz}
Octahedron	6.00	6.00	−4.00	−4.00	−4.00
Trigonal bipyramid (T.P.)	−0.82	7.07	−0.82	−2.72	−2.72
Square pyramid (S.P.)	9.14	0.86	−0.86	−4.57	−4.57
Pentagonal bipyramid (P.P.)	2.82	4.93	2.82	−5.28	−5.28
Octahedral wedge (O.W.)	8.79	1.39	−1.51	−2.60	−6.08
Square planar	12.28	−4.28	2.28	−5.14	−5.14

Table 3.15. Crystal Field Activation Energies for Two Transition States with High-Spin Configurations

	CFSE			CFAE	
No. d electrons	Octahedron	S. P.	O. W.	S. P.	O. W.
0, 10	0.0	0.0	0.0	0.0	0.0
1, 6	4.00	4.57	6.08	−0.57	−2.08
2, 7	8.00	9.14	8.68	−1.14	−0.68
3, 8	12.00	10.00	10.20	2.00	1.80
4, 9	6.00	9.14	8.79	−3.14	−2.79
5, 10	0.0	0.0	0.0	0.0	0.0

However, an octahedral wedge gives a more favorable transition state for the (d^1, d^6) and (d^3, d^8) configurations and a pentagonal bipyramid is predicted to be the best for the (d^2, d^7) configurations with CFAE = −2.56 Dq.

Table 3.16 summarizes the predictions and results for the 2+ metal ions of the first transition series. The theory correctly predicts that nickel(II) and vanadium(II) should have the smallest exchange rates and that chromium(II) and copper(II) should have the largest. The next most labile are predicted to be cobalt(II) and iron(II), but they are actually less labile than manganese(II) and probably zinc(II).

The predictions of absolute and relative activation energies are less successful. For example, $Ni(OH_2)_6^{2+}$ has Dq = 850 cm^{-1} = 10 kJ and $V(OH_2)_6^{2+}$ has Dq = 1240 cm^{-1} = 14.8 kJ, so that their ΔH^* values are calculated to be 18 and 26.6 kJ mol^{-1}, respectively. The differences from

Table 3.16. Water Solvent Exchange Rates (25°C), Activation Parameters, and Predicted Crystal Field Activation Energies for First-Row Transition-Metal Ions

		CFAE	k (s^{-1})	ΔH^* (kJ mol^{-1})	ΔS^* (J mol^{-1} K^{-1})	ΔV^* (cm^3 mol^{-1})
$V(OH_2)_6^{2+}$	d^3	1.80 (O.W.)	87	61.8	−0.4	−4.1
$Cr(OH_2)_6^{2+}$	d^4	−3.14 (S.P.)	>10^8			
$Mn(OH_2)_6^{2+}$	d^5	0.0	2.1 x 10^7	32.9	5.7	−5.4
$Fe(OH_2)_6^{2+}$	d^6	−2.08 (O.W.)	4.4 x 10^6	41.4	21.2	3.8
$Co(OH_2)_6^{2+}$	d^7	−2.56 (P.P.)	3.2 x 10^6	46.9	37.2	6.1
$Ni(OH_2)_6^{2+}$	d^8	1.80 (O.W.)	3.2 x 10^4	56.9	32.0	7.2
$Cu(OH_2)_6^{2+}$	d^9	−3.14 (S.P.)	>10^7			
$Zn(OH_2)_6^{2+}$	d^{10}	0.0	>10^7			

the experimental numbers could be ascribed to solvation effects. The predicted difference of 8.6 kJ mol^{-1} in the ΔH^* values is larger than the observed value of 5.1 kJ mol^{-1}, but one must allow for a ± 2 kJ mol^{-1} error in the experimental values. The larger ΔH^* for the 3+ ions would be attributed to differences in Dq. For chromium(III), the Dq = 2000 cm^{-1} = 23.7 kJ, so that the calculated $\Delta H^* = 42.7$ kJ mol^{-1}, and the difference between this value and that for nickel(II) is predicted to be 24.7 kJ mol^{-1} compared to the observed value of 53.1 kJ mol^{-1}. Again, solvation differences with different charge types may explain this discrepancy.

The general impression is that the crystal field approach has pointed out a significant feature for the understanding of the reactivity differences. However, this is not the whole story, and it is probably too crude an approximation in the form used by Basolo and Pearson to be capable of assigning small differences or preferred geometries for transition states.

There have been attempts to refine the theory. Breitschwerdt[95] allowed the effective charge on the metal ion to change with the number of ligands, so that Dq varied between the ground and transition states, and Dq also decreased by 40 percent for ligands in the plane of the pentagonal bipyramid compared to the axial ligands. It was found that a square pyramid was the most stable transition state for all the 2+ ions and that all the CFAEs are positive, as shown in Table 3.17.

Spees et al.[96] presented a more extensively parameterized ligand field model and included the possibility of a change of spin state in the "intermediate." However, this approach seems to have been no more effective than earlier versions with regard to the water exchange reactions.

3.7.a.iii   Other Theoretical Applications

Burdett[97] has applied the angular overlap bonding model to this problem. This is essentially an extended Hückel molecular orbital approach. The change in bonding energy between the octahedral ground state and a square pyramid transition state was calculated in terms of the exchange integral (β) and the overlap integral (S). It was also argued that βS^2 would increase with atomic number across the transition series. The energy loss for d^0 to d^3 is $2\beta S^2$, for d^4 to d^8 is $1\beta S^2$ and for d^9 and d^{10} is zero. This predicts the abrupt change in reactivity that is observed between d^3 and d^4 and between d^8 and d^9, but further quantitative predictions are not possible, nor were other transition states considered. The angular overlap model has been applied by Mønsted[98] to the aquation reactions of $(H_2O)_5Cr^{III}X$ and $(NH_3)_5Cr^{III}X$ complexes, with the conclusion that an I_a octahedral wedge "transition state" is preferred.

A more sophisticated quantum mechanical model has been applied by Rode et al.[99] to calculate the hydration energies of these metal ions, by including effects from two hydration shells beyond the first coordination sphere. The stabilization per water molecule in the first coordination sphere, $\Delta E(I)$,

Table 3.17. Modified Crystal Field Activation Energies

No. d electrons	CFAE (S.P.)
0, 5	2.80 Dq
1, 6	2.87 Dq
2, 7	3.65 Dq
3, 8	4.80 Dq
4, 9	1.06 Dq

was calculated and was found to correlate reasonably well with the water exchange rates [i.e., $\Delta G^*(25°C)$], but the absolute energies are different by a factor of about seven. To account for this discrepancy, Rode et al. propose that there is synchronous movement of the leaving and entering groups between the first and second coordination spheres, so that there is bond making as well as bond breaking. This is taken into account in Eq. (3.61), where $\Delta E(II)$ is the stabilization energy of a water molecule in the second coordination sphere and κ is an adjustable parameter that varies with the degree of movement between the two solvent shells:

$$\Delta G^* = \kappa [\Delta E(I) + \Delta E(II)] - \Delta E(I) \qquad (3.61)$$

The parameters in kJ mol^{-1} for three 2+ transition-metal ions are given in Table 3.18. These numbers illustrate an inevitable feature of this type of calculation in that one is taking the difference between two large numbers of uncertain validity in order to obtain the parameter of relevance for the kinetic interpretation. The preceding calculations should really yield ΔH^*, but Rode et al. argue that this does not matter because the ΔS^* differences are small; this is only marginally true. Of course, the adjustable parameter κ could be used to fit the ΔH^* values.

Connick and Alder[100] have applied molecular modeling to attempt to understand the nature of this exchange process, and more of this type of approach is likely to appear in the future as these methods become more available and more sophisticated.

Table 3.18. Solvation Energy Parameters for Eq. (3.61)

Metal Ion	$\Delta E(I)$	$\Delta E(II)$	κ	ΔG^*_{cal}	ΔG^*_{obs}	ΔH^*_{obs}
Fe^{2+}	291	87	0.87	36.8	38.7	41.4
Co^{2+}	313	94	0.87	41.0	40.4	46.9
Ni^{2+}	324	90	0.90	48.6	48.3	56.9

It has been noted that the ΔV^* values for the 2+ ions of the first transition series become increasingly positive with atomic number. Merbach has suggested that this represents a trend from I_a for $V(OH_2)_6^{2+}$ to I_d for $Ni(OH_2)_6^{2+}$, and that the reason for this is the decreasing size of the metal ion. The ΔV^* values for the $M(OH_2)_5(OH)^{2+}$ ions are more positive than those for the corresponding $M(OH_2)_6^{3+}$ ions [M = Cr(III), Fe(III), and Ru(III)], suggesting that the former species have more dissociative character. Swaddle has noted that the partial molar volumes of $M(OH_2)_n^{z+}$ ions can be calculated from

$$\overline{V}^o_{abs} = 2.523 \times 10^{-6}\left(r_M + \Delta r\right) - 18.07\, n - \frac{417.5\, z}{\left(r_M + \Delta r\right)} \qquad (3.62)$$

where $\overline{V}^o_{abs}$ is the absolute volume relative to $\overline{V}^o_{abs}(H^+) = -5.4$ cm^3 mol^{-1}, r_M is the ionic radius of the metal ion (pm), Δr is the apparent radius of the coordinated water molecule determined empirically, and 18.07 is the partial molar volume of water; the last term accounts for solvent electrostriction around the charged ion. Swaddle[101] has used this equation, with appropriately adjusted values of r and n, to estimate a limiting value of $\Delta V^* \approx 13$ cm^3 mol^{-1} for a **D** mechanism and $\Delta V^* \approx -13$ cm^3 mol^{-1} for an **A** mechanism. This implies that $Ti(OH_2)_6^{3+}$ has a mechanism close to the **A** limit.

3.7.b   Solvent Exchange in Nonaqueous Solvents

Although water is the solvent of most general interest, there has been a great deal of work in other solvents, such as acetonitrile, dimethylsulfoxide, methanol, and *N,N*-dimethylformamide. In general, the purpose is to gain some understanding of the effect of ligand size, basicity, and so on, on the solvent exchange rate. The main complication is that both the bulk solvent and the exchanging ligand are changed at the same time, so that individual factors affecting the exchange rate are difficult to separate. Some representative results are given in Table 3.19.

The reactivity pattern observed in water generally is maintained in other solvents. Nickel(II) always shows substantially smaller exchange rate constants and higher ΔH^* values, and trends in ΔV^* with atomic number are also maintained.

There are some disturbing features of the activation parameters with regard to the conventional interpretations of the data in water. For example, the order of Dq values for the solvents is $NH_3 > CH_3CN > DMF > H_2O \approx CH_3OH$. If crystal field effects are the determining factor for the ΔH^* values, then one should expect these to be in the same order for the various solvents. In fact, the order of ΔH^* values for nickel(II) is $CH_3OH > DMF > CH_3CN \approx NH_3 > H_2O$.

Table 3.19. Nonaqueous Solvent Exchange Rate Constants (25°C), Enthalpies, Entropies, and Volumes of Activation[a]

Metal	Solvent	k (s^{-1})	ΔH^* $(kJ\ mol^{-1})$	ΔS^* $(J\ mol^{-1}\ K^{-1})$	ΔV^* $(cm^3\ mol^{-1})$
$Cr^{2+\ b}$	CH_3OH	1.2×10^8	31.6	16.6	
Mn^{2+}	CH_3OH	3.7×10^5	25.9	–50.2	–5.0
Fe^{2+}	CH_3OH	5.0×10^4	50.2	12.6	0.4
Co^{2+}	CH_3OH	1.8×10^4	57.7	30.1	8.9
Ni^{2+}	CH_3OH	1.0×10^3	66.1	33.5	11.4
Cu^{2+}	CH_3OH	3.1×10^7	17.2	–44.0	8.3
Mn^{2+}	CH_3CN	1.4×10^7	29.6	–8.9	–7.0
Fe^{2+}	CH_3CN	6.6×10^5	41.4	5.5	3.0
Co^{2+}	CH_3CN	3.4×10^5	49.5	27.1	9.9
Ni^{2+}	CH_3CN	3.1×10^3	60.8	25.8	7.3
Co^{2+}	NH_3	5.0×10^7	45.8	31.2	
Ni^{2+}	NH_3	7.0×10^4	57.3	40.2	5.9
Mn^{2+}	$(CH_3)_2NCHO$	2.2×10^6	34.6	–7.4	2.4
Fe^{2+}	$(CH_3)_2NCHO$	9.7×10^5	43.0	13.8	8.5
Co^{2+}	$(CH_3)_2NCHO$	3.9×10^5	56.9	52.7	6.7
Ni^{2+}	$(CH_3)_2NCHO$	3.8×10^3	62.8	33.5	9.1
Fe^{3+}	$(CH_3)_2NCHO$	6.1×10^1	42.3	12.1	–5.4
Cr^{3+}	$(CH_3)_2NCHO$	3.3×10^{-7}	97.1	–43.5	–6.3

[a] Original references in: Merbach, A. E. *Pure Appl. Chem.* **1982**, *54*, 1479; Ibid. **1987**, *59*, 161; Rusnak, L. L.; Yang, E. S.; Jordan, R. B. *Inorg. Chem.* **1978**, *17*, 1810.
[b] Li, C.; Jordan, R. B. *Inorg. Chem.* **1987**, *26*, 3855.

Furthermore, the ΔV^* values are almost invariant with changes in solvent for a given metal ion. If the mechanism is I_d for nickel(II), as is commonly assumed, then one might expect the ΔV^* values to parallel the size of the leaving group (solvent), based on a straightforward extension of Eq. (3.62) from water to other solvents. Yet the ΔV^* for DMF is only 2 cm^3 mol^{-1} larger than that for water, although the partial molar volumes of the solvents are 115.4 and 18 cm^3 mol^{-1}, respectively.

The general interpretation of these results is that there are specific solvation effects operating in different solvents, and these are not taken into account by any of the simple models. However, it is not widely acknowledged that this greatly weakens all of the more simplistic rationalizations that are used to explain the results of these types of studies.

It was suggested by Jordan and co-workers[102] that the ΔH^* values could be correlated by Eq (3.63), which involves a crystal field parameter (a_M)

Table 3.20. Correlation of ΔH^* Values (kcal mol^{-1}) for Solvent Exchange[a]

Solvent	Dq	b_S	Ni(II)	Co(II)	Fe(II)	Mn(II)
NH$_3$	3.10	1.63	11.0 (14.8)[b]	11.2 (11.9)	(10.4)	8.0 (7.3)
CH$_3$CN	3.03	1.52	15.0 (14.4)	11.4 (11.5)	9.7 (10.1)	7.3 (7.0)
DMF	2.50	4.64	15.0 (15.3)	13.6 (12.9)	11.7 (11.7)	8.9 (9.1)
H$_2$O	2.46	3.58	14.4 (14.1)	11.9 (11.7)	7.7 (10.5)[b]	7.9 (8.0)
CH$_3$OH	2.43	5.44	15.8 (15.8)	13.8 (13.5)	12.0 (12.3)	6.2 (9.8)[b]
DMSO	2.25	4.12	13.0 (13.7)	12.2 (11.6)	11.3 (10.5)	7.4 (8.2)
c_M			4.26	3.31	2.82	1.80
a_M			4.26	3.16	2.51	1.88

[a] Values in parentheses are predicted from the correlation.
[b] These values were not used to obtain the correlation parameters.

and a solvent parameter (b_S). The latter would be constant for a particular solvent with 2+ ions and the former would depend on the metal ion.

$$\Delta H^* = a_M\, g_M\, (\, Dq_{Ni}\,) + b_S = c_M\, (\, Dq_{Ni}\,) + b_S \qquad (3.63)$$

The g_M is obtained from spectroscopic data that gives the proportionality between the widely available Dq values for nickel(II) and those of the other metal ions. The results of the 1978 correlation are shown in Table 3.20.

Since the original correlation, a number of experimental values have been redetermined and the best current values for these are Ni(II)–NH$_3$, 13.7; Co(II)–CH$_3$CN, 11.7; Fe(II)–CH$_3$CN, 9.9; Fe(II)–H$_2$O, 9.9. The newer values are equally or more consistent with the predictions of this correlation. It remains to be determined whether the Mn(II)–CH$_3$OH data are in error. Although this correlation seems to have some valuable predictive property, there is as yet no rationalization of the b_S values in terms of any solvent property.

3.8 LIGAND SUBSTITUTION ON LABILE TRANSITION-METAL IONS

The common complexes of the M(II) and M(III) transition-metal ions of the first transition series are labile, except for Cr(III) and low-spin Co(III), Fe(II), and Fe(III). These labile ions form a wide range of complexes of general chemical and biochemical importance. As a result, there have been many studies of the kinetics and equilibrium constants for reactions of the general form

$$M\,(Solvent)_6{}^{n+} + L^{z-} \;\rightleftharpoons\; M\,(Solvent)_5(L)^{(n-z)+} + Solvent \qquad (3.64)$$

The majority of this work is in water, but there are an increasing number of studies in nonaqueous solvents. The results have been the subject of numerous reviews.[103]

3.8.a General Reactivity Trends

The rate constants for these reactions are normally of the same order of magnitude as those for solvent exchange. As a result, the reactions typically have half-times in the microsecond to second range at normal concentrations. Therefore, the experimental techniques are stopped-flow and various relaxation methods pioneered by Eigen (e.g., T-jump, P-jump).[104] The lability of these systems means that one cannot do the type of competition studies that rely on product analysis to yield mechanistic information. Rather, the mechanistic implications of these studies are primarily concerned with the evaluation of entering group effects. Since the leaving group is the solvent, it cannot be systematically changed without introducing substantial changes to the solvation energies of the species involved.

The reactions of Ni(II) are slower than those of the other M(II) ions in this group, and this metal ion has been studied most extensively because the rates fall conveniently in the range of the stopped-flow technique. The substitution kinetic results on $Ni(OH_2)_6^{2+}$ have been reviewed by Wilkins.[105] Some representative results are given in Table 3.21. The first impression of these data is that there are significant entering group effects on the rate constant, so that the mechanism appears to be I_a. However, there is no correlation with the nucleophilicity of L as judged by the pK_a values of the entering ligands. An analysis of a more extensive set of this type of data led Wilkins and Eigen[106] to suggest that the apparent rate constant was strongly correlated with the charge on L. Within modest limits, the rate constants (M^{-1} s^{-1}) are ~3 x 10^5 for L^{2-}, ~1 x 10^4 for L^{1-}, ~3 x 10^3 for L^0, and ~500 for L^{1+}. These observations led Eigen and Wilkins to propose what is now the classic mechanism for this system, called the *Dissociative ion pair mechanism* or the *Eigen–Wilkins mechanism*; it is formulated in Scheme 3.4.

Table 3.21. Rate Constants (25°C) for Some Substitution Reactions on $Ni(OH_2)_6^{2+}$

Entering Group (L)	pK_a	k (M^{-1} s^{-1})
$H_3CPO_4^{2-}$	~2	2.9 x 10^5
F^-	~3	8.0 x 10^3
NH_3	9.2	4.5 x 10^3
NH_2CH_3	10	1.3 x 10^3
$NH_2(CH_2)N(CH_3)_3^+$	7	4.0 x 10^2

Scheme 3.4

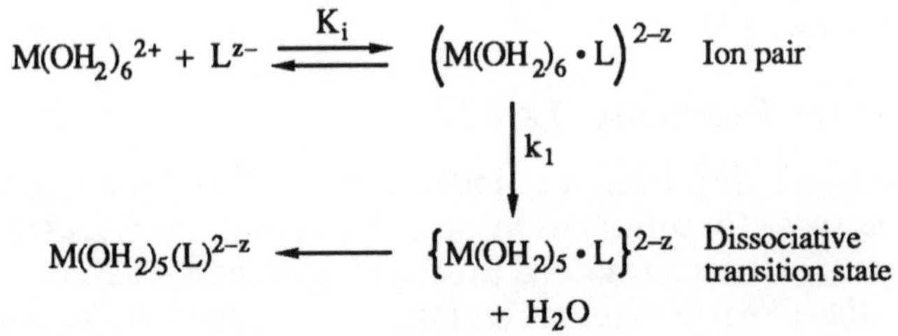

$$M(OH_2)_6{}^{2+} + L^{z-} \; \underset{\longleftarrow}{\overset{K_i}{\rightleftharpoons}} \; \left(M(OH_2)_6 \cdot L\right)^{2-z} \quad \text{Ion pair}$$

$$\downarrow k_1$$

$$M(OH_2)_5(L)^{2-z} \longleftarrow \left\{M(OH_2)_5 \cdot L\right\}^{2-z} \quad \begin{array}{l}\text{Dissociative}\\ \text{transition state}\end{array}$$
$$+ \; H_2O$$

The rate law for this system with a rapid pre-equilibrium has been developed previously. These studies are usually carried out with total [M] >> [L], in order to prevent formation of higher complexes such as $M(OH_2)_4(L)_2$; therefore, the experimental pseudo-first-order rate constant is given by

$$k_{exp} = \frac{k_1 \, K_i}{1 + K_i \, [M(OH_2)_6{}^{2+}]} \tag{3.65}$$

The theoretical expression for K_i, discussed on page 33, was developed in part to analyze these results. When L is neutral or uninegative, K_i is predicted to be ≤ 1, and for most studies total $[M] \approx 10^{-2}$ M, so that $K_i[M(OH_2)_6{}^{2+}] \ll 1$. Under these conditions, Eq. (3.65) reduces to

$$k_{exp} = k_1 \, K_i \tag{3.66}$$

Since K_i depends on the charge of L, this expression explains the variation of k_{exp} with the charge of the entering group. In many interpretations of experimental results, k_{exp} is divided by a calculated value of K_i in order to obtain k_1, and then k_1 is compared to the rate constant for solvent exchange, which also is assumed to be dissociatively activated. In many instances, this analysis works well. In general, since K_i is of the order of magnitude of 1, the value of k_{exp} is quite similar to that for solvent exchange.

For Mn(II), Fe(II), and Co(II) the preceding type of analysis also indicates a lack of entering group effects, and the mechanism appears to be I_d. However, Merbach and co-workers[107] have interpreted the changes in ΔV^* to indicate that there is a mechanistic trend across the first transition series from I_a for V(II) to I_d for Ni(II).

The best case for an I_a mechanism from entering group effects comes from the results of Diebler and co-workers[108] for $Ti(OH_2)_6{}^{3+}$, as given in Table 3.22. These results do not correlate simply with the charge of the entering group, but do parallel the nucleophilicity as measured by the pK_a. These reactions are still presumed to proceed through an ion pair, but the rate-controlling step has the entering group dependence expected for an I_a mechanism. Merbach has suggested that the mechanism is **A** on the basis of the ΔV^* for the water exchange.[109]

Table 3.22. Rate Constants (15°C) for Substitution Reactions on $Ti(OH_2)_6^{3+}$

Entering Group	pK_a	k_{exp} (M^{-1} s^{-1})
$ClCH_2CO_2H$		6.7×10^2
CH_3CO_2H		9.7×10^2
H_2O	-1.74	8.6×10^3
NCS^- [a]	-1.84	8.0×10^3
$HO_2CCO_2^-$	1.23	3.9×10^5
$Cl_2CHCO_2^-$	1.25	1.1×10^5
$HO_2CCH_2CO_2^-$	2.43	4.2×10^5
$ClCH_2CO_2^-$	2.46	2.1×10^5
$HO_2CCH(CH_3)CO_2^-$	2.62	3.2×10^5
$CH_3CO_2^-$	4.47	1.8×10^6

[a] At 8-9°C, Diebler, H. *Z. Phys. Chem.* **1969**, *68*, 64.

Substitution reactions on vanadium(III) have been of long-standing interest since Taube's prediction that such a d^2 system should be labile with an **A** mechanism. Unfortunately, these studies are difficult because $V(OH_2)_6^{3+}$ is readily oxidized by air and perchlorate ion. The results of several studies as summarized by Patel and Diebler[110] are given in Table 3.23. These rate constants vary widely with the entering group and do not show the correlation with charge expected for a dissociative ion pair mechanism. These observations, together with the negative ΔS^* and ΔV^* for water exchange, are taken as evidence for an I_a mechanism, confirming Taube's prediction.

3.8.b   Substitution on Iron(III) and the Proton Ambiguity

Despite its common occurrence and importance, $Fe(OH_2)_6^{3+}$ was the last of the air-stable first-row transition-metal ions to have its water exchange rate

Table 3.23. Rate Constants (25°C) for Substitution Reactions on $V(OH_2)_6^{3+}$

Entering Group	k (M^{-1} s^{-1})
$HC_2O_4^-$	1.3×10^3
SCN^-	1.1×10^2
Br^-	≤ 10
Cl^-	≤ 3
HN_3	0.4
H_2O	5.0×10^2 (s^{-1})

determined. The difficulty is the tendency of $Fe(OH_2)_6^{3+}$ to hydrolyze and polymerize in dilute aqueous acid, forming $Fe(OH_2)_5(OH)^{2+}$ and $(H_2O)_4Fe(OH)_2Fe(OH_2)_4^{4+}$, respectively, as the major species. There have been numerous kinetic studies of substitution on $Fe(OH_2)_6^{3+}$, but the results were difficult to interpret because of the lack of a water exchange rate until 1981 and because of the proton ambiguity discussed later.

For the typical system in which the entering group has an ionizable proton, and for iron(III) concentrations sufficiently low to avoid the hydrolyzed dimer, the rapid equilibria involving the reactants are represented by

$$
\begin{aligned}
Fe(OH_2)_6^{3+} &\xrightleftharpoons{K_m} (H_2O)_5Fe(OH)^{2+} + H^+ \\
HL &\xrightleftharpoons{K_a} L^- + H^+
\end{aligned}
\tag{3.67}
$$

where $K_m = 1.6 \times 10^{-3}$ M (25°C). In the following development the charges of the iron and ligand species have been omitted. The total iron(III) concentration in the reactants is given by

$$[Fe]_{tot} = [FeOH_2] + [FeOH] \tag{3.68}$$

From Eqs. (3.67) and (3.68), the following equations for the concentrations of the iron(III) species in terms of $[Fe]_{tot}$ can easily be developed:

$$[FeOH_2] = \frac{[H^+]}{K_m + [H^+]}[Fe]_{tot} \quad \text{and} \quad [FeOH] = \frac{K_m}{K_m + [H^+]}[Fe]_{tot} \tag{3.69}$$

Similarly, since the total entering group concentration is given by

$$[L]_{tot} = [HL] + [L] \tag{3.70}$$

the concentrations of the ligand species in terms of $[L]_{tot}$ are

$$[HL] = \frac{[H^+]}{K_a + [H^+]}[L]_{tot} \quad \text{and} \quad [L] = \frac{K_a}{K_a + [H^+]}[L]_{tot} \tag{3.71}$$

The possible substitution reactions in this system are given by

$$
\begin{aligned}
Fe(OH)^{2+} + L &\xrightarrow{k_1} \\
Fe(OH)^{2+} + HL &\xrightarrow{k_2} \\
Fe(OH_2)^{3+} + L &\xrightarrow{k_3} \\
Fe(OH_2)^{3+} + HL &\xrightarrow{k_4}
\end{aligned}
\tag{3.72}
$$

The rate of formation of product (P) is given by Eq. (3.73), so that substitution from Eqs. (3.68 to 3.71) into Eq. (3.73) leads to Eq. (3.74):

$$\frac{d[P]}{dt} = (k_1[L] + k_2[HL])[FeOH] + (k_3[L] + k_4[HL])[FeOH_2] \quad (3.73)$$

$$= \left(\frac{k_1 K_m K_a + (k_2 K_m + k_3 K_a)[H^+] + k_4[H^+]^2}{(K_m + [H^+])(K_a + [H^+])} \right) [Fe]_{tot}[L]_{tot} \quad (3.74)$$

which can be simplified to

$$\frac{d[P]}{dt} = k_{exp}[Fe]_{tot}[L]_{tot} \quad (3.75)$$

In Eq. (3.75) it has been assumed that the experimental conditions are such that $[H^+]$ is constant for a particular kinetic run. The overall experimental study will involve determining k_{exp} at different $[H^+]$ and fitting this information to the $[H^+]$ dependence of k_{exp} predicted by Eq. (3.74).

For most such studies, $[H^+] >> K_m (= 1.6 \times 10^{-3}$ M), so that two limiting conditions for k_{exp} can be identified depending on the strength of the acid HL.

1. *Strong acid ligands*: $K_a >> [H^+]$ and

$$k_{exp} = \frac{k_1 K_m}{[H^+]} + \frac{k_2 K_m}{K_a} + k_3 + \frac{k_4[H^+]}{K_a} \quad (3.76)$$

2. *Weak acid ligands*: $[H^+] >> K_a$ and

$$k_{exp} = \frac{k_1 K_m K_a}{[H^+]^2} + \frac{k_2 K_m}{[H^+]} + \frac{k_3 K_a}{[H^+]} + k_4 \quad (3.77)$$

The *proton ambiguity* refers to the fact that the $[H^+]$ dependence of k_{exp} does not allow one to separate $k_2 K_m / K_a$ from k_3 in the first case, or $k_2 K_m$ from $k_3 K_a$ in the second case. This results because both transition states, Fe(OH) • HL or Fe(OH$_2$) • L contain one ionizable proton that is in different sites. *The kinetics can only give the composition of the transition state, not its structure.*

The experimental pseudo-first-order rate constant for almost all substitution reactions on aqueous iron(III) has the general form of

$$k_{exp} = k' + k''[H^+]^{-1} \quad (3.78)$$

The specific assignment of k' and k" can be made by comparison to the theoretical expressions for the appropriate case, except for the proton ambiguity problem.

1. *Strong acid ligands*: ($K_a \gg$ [H^+]). A comparison of Eqs. (3.76) and (3.78) shows that k" = $k_1 K_m$, and k_1 can be calculated since K_m is known. However, k' = $k_3 + k_2 K_m / K_a$ and cannot be uniquely assigned except for very strong acid ligands, such as Cl^- and Br^- ($K_a \approx 10^8$ M). For such ions, the large K_a makes the second term negligible even if k_2 has a diffusion limiting value of ~10^{10} M^{-1} s^{-1}.

2. *Weak acid ligands*: ([H^+] $\gg K_a$). From Eqs. (3.77) and (3.78), k' = k_4 and k" = $k_2 K_m + k_3 K_a$. If K_a is very small (e.g., <10^{-10} M), it may be possible to eliminate the $k_3 K_a$ term again on the grounds that the k_3 would need to be beyond the diffusion-controlled limit.

The usual strategy in this area has been to calculate both k_2 and k_3 in ambiguous cases, and then compare k_1 and k_2, assuming that they should be similar for a dissociative mechanism, and likewise for k_4 and k_3. Until the water exchange rates were determined, this strategy led to rather ambiguous mechanistic conclusions.[111,112] The current interpretation is that $Fe(OH_2)_6^{3+}$ has an I_a mechanism [like Ti(III) and V(III)], whereas $(H_2O)_5Fe(OH)^{2+}$ has an I_d mechanism. Some of the kinetic data leading to this conclusion are given in Table 3.24. In the case of 4-isopropyltropolone (Hipt), the assignment to k_3 was made because the alternative assignment to k_2 (FeOH^{2+} + H$_2$ipt$^+$) gives a larger value for k_2 than k_1 (FeOH^{2+} + Hipt) and would be inconsistent with expected charge effects.

Table 3.24. Rate Constants[a] (25°C) for Substitution Reactions on Aqueous Iron(III)

Entering Group	k (M^{-1} s^{-1})	
	$(H_2O)_5Fe(OH)^{2+}$	$Fe(OH_2)_6^{3+}$
SO_4^{2-}	1.1×10^5	2.3×10^3
Cl^-	5.5×10^3	4.8
Br^-	2.8×10^3	1.6
NCS^-	5.1×10^3	9.0×10^1
$Cl_3CCO_2^-$	7.8×10^3	6.3×10^1
$Cl_2HCCO_2^-$	1.9×10^4	1.2×10^2
$ClH_2CCO_2^-$	4.1×10^4	1.5×10^3
H_3CCO_2H	$\leq 2.8 \times 10^3$	2.7×10^1
C_6H_5OH	1.5×10^3	
$C_{10}H_{12}(=O)(OH)$[b]	6.3×10^3	2.2×10^1
$H_3CC(O)NH(OH)$	2.0×10^3	1.2
H_2O (in s^{-1})	1.2×10^5	1.6×10^2

[a] Data from Grant, M.; Jordan, R. B. *Inorg. Chem.* **1981**, *20*, 55, where original references are given.

[b] 4-isopropyltropolone, Ishihara, K.; Funahashi, S.; Tanaka, M. *Inorg. Chem.* **1983**, *22*, 194.

3.9 KINETICS OF CHELATE FORMATION

Complexes containing bidentate, tridentate, and so on, ligands are very common with metal ions. They are especially important because of their apparently enhanced stability, often referred to as the "chelate effect." Such ligands have two or more potential donor atoms separated by two, three, or four other atoms in the molecule. Some common examples are oxalate, glycine, salicylate, ethylenediamine, acetylacetonate, 2,2'-bipyridyl, and ethylenediaminetetraacetic acid.

3.9.a Rate-Controlling Step

The formation of these complexes from the solvated metal ion involves the displacement of two solvent molecules for a bidentate chelate. The process might be viewed as being completely concerted, with simultaneous replacement, or as a stepwise process. The concerted process seems highly improbable and is generally not considered as a mechanistic possibility. The individual steps that are usually considered in the stepwise process are shown in Scheme 3.5 for glycine as an example.

Scheme 3.5

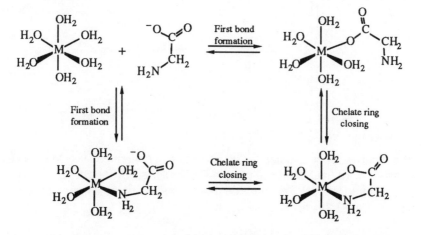

This scheme shows two mechanistic complications that can occur in chelate formation reactions. First of all, there are two potential rate-controlling steps, either *first bond formation* or *chelate ring closing*. Second, for unsymmetrical chelates such as glycine, there is an ambiguity as to which donor group may coordinate first. The latter is not a complication for symmetrical chelates such as oxalate or ethylenediamine.

Prior to about 1970, it was commonly assumed that the chelate ring-closing step was very fast relative to the first bond formation. The rationale for this assumption was that the free end of the chelate in the monodentate intermediate would have a very high effective concentration in the neighborhood of the metal ion so that ring closing would be fast. However, there is increasing evidence that substitution reactions on the M(II) first row transition-metal ions have considerable dissociative character, and the dissociative ion pair mechanism has achieved wider acceptance. If the mechanism is dissociative, then the rate-limiting feature is the breaking of the M—OH$_2$ bond, and the effective concentration of the entering group is no greater than it would be in the ion pair. The idea that first bond formation is always rate limiting still persists in some quarters, perhaps because the experimental rate constants for chelate formation are often quite similar to those for monodentate complex formation with a given metal ion.

A proper consideration of the kinetic situation requires an analysis of the theoretical rate law for such a system. The following analysis is somewhat simplified in that a symmetrical bidentate chelate is assumed, and the ring closing is not reversible. A complete version has been given by Letter and Jordan.[113] These reactions are usually done in dilute acid (pH 5–7), so that protonation of the entering group has been included in Scheme 3.6, but charges and coordinated solvent have been omitted.

Scheme 3.6

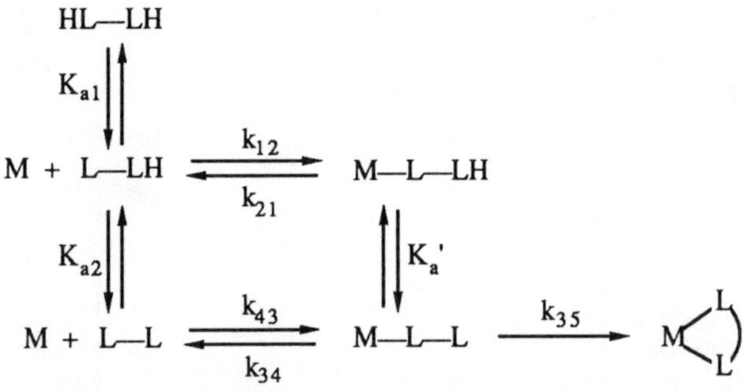

If a steady state is assumed for the monodentate intermediates (M—L—LH and M—L—L), and the usual pseudo-first-order conditions of [M] >> [L]$_{tot}$ and constant [H$^+$] are assumed, then the pseudo-first-order rate constant is given by Eq. (3.79) for K$_{a1}$ >> [H$^+$]:

$$k_{exp} = \frac{\left(\dfrac{k_{12}\,[H^+] + k_{43}\,K_{a2}}{K_{a2} + [H^+]} \right) k_{35}\,K_a'}{k_{21}\,[H^+] + K_a'(\,k_{34} + k_{35}\,)}\,[M] \qquad (3.79)$$

This equation can be rearranged to give Eq. (3.80), which has the form typically used to analyze data by plotting the left-hand side versus [H$^+$]$^{-1}$:

$$\frac{k_{exp}}{[M]}\left(\frac{K_{a2} + [H^+]}{[H^+]} \right) = \frac{\left(k_{12} + k_{43}\,K_{a2}\,[H^+]^{-1} \right) k_{35}\,K_a'}{k_{21}\,[H^+] + K_a'(\,k_{34} + k_{35}\,)} \qquad (3.80)$$

In a typical study, such as that of Wilkins and co-workers,[114] this plot is found to be linear. The traditional explanation for this is that $K_a'k_{35} \gg k_{21}[H^+]$ and $k_{35} \gg k_{34}$, so that Eq. (3.80) simplifies to Eq. (3.81), which will give k_{12} as the intercept and $k_{43}K_{a2}$ as the slope:

$$\frac{k_{exp}}{[M]}\left(\frac{K_{a2} + [H^+]}{[H^+]}\right) = k_{12} + k_{43}\, K_{a2}\, [H^+]^{-1} \tag{3.81}$$

This rate law is consistent with the observations, but there is at least one notable problem. For amino acids such as glycine ($HO_2CCH_2NH_3^+$), it is always found that $k_{12} = 0$ (i.e., there is no significant intercept for the plot). This implies that the zwitterion ($^-O_2CCH_2NH_3^+$) is unreactive. This lack of reactivity always seemed remarkable but was attributed to intramolecular hydrogen bonding between the amino and carboxylate groups. However, monoprotonated ethylenediamine ($H_2NCH_2CH_2NH_3^+$) does react, although the charge is less favorable and there is a greater likelihood of hydrogen bonding.

These and other problems caused Letter and Jordan to reconsider the assumptions used to reduce the complete rate law. Is it reasonable that $K_a'k_{35} \gg k_{21}[H^+]$? If one uses the nickel(II)–glycine system for analysis, then the value of k_{21} can be estimated from the corresponding acetate system, so that $k_{21} \approx 10^4\ s^{-1}$ and the K_a of $((NH_3)_5Co—O_2CCH_2NH_3^+)^{3+}$ can be used to estimate that $K_a' \approx 3 \times 10^{-10}\ M$. Then, the condition that $K_a'k_{35} \gg k_{21}[H^+]$ at pH 6 requires that $k_{35} \geq 10^8\ s^{-1}$. However, water exchange on nickel(II) has $k = 3 \times 10^4\ s^{-1}$, and all the available evidence indicates an I_d mechanism for substitution on Ni(II). Therefore, the preceding condition requires that the water in the monodentate intermediate has been labilized by about 10^4, an effect without precedent.

On the other hand, if one assumes that $k_{21}[H^+] \gg K_a'k_{35}$, then the denominator in the complete rate law can be rearranged using the principle of detailed balancing, which shows that $K_a'(k_{12}/k_{21}) = K_{a2}(k_{43}/k_{34})$, to give

$$k_{21}\,[H^+] + (k_{34} + k_{35})\,K_a' = \frac{k_{21}\,[H^+]}{k_{12}}\left(k_{12} + \frac{k_{43}\,K_{a2}\,k_{35}}{k_{34}\,[H^+]}\right) \tag{3.82}$$

A comparison of the right- and left-hand sides of Eq. (3.82), combined with the assumed inequality, requires that $k_{12} \gg k_{43}K_{a2}k_{35}/k_{34}[H^+]$. Then, if $k_{35} \gg k_{34}$, as assumed before, $k_{12} \gg k_{43}K_{a2}/[H^+]$ and the rearranged version of Eq. (3.79) simplifies to

$$\frac{k_{exp}}{[M]}\left(\frac{K_{a2} + [H^+]}{[H^+]}\right) = \frac{k_{12}\,k_{35}\,K_a'}{k_{21}\,[H^+]} = \frac{k_{43}\,k_{35}\,K_{a2}}{k_{34}\,[H^+]} \tag{3.83}$$

This equation is consistent with the experimental observations; a zero intercept is required but does not imply that k_{12} is zero. This rate law corresponds to the condition of a rapid pre-equilibrium formation of the monodentate species (k_{43}/k_{34}) followed by *rate-controlling chelate ring closure*. The values originally associated with k_{43} are actually $k_{43}k_{35}/k_{34}$. Since these are numerically similar to rate constants observed for monodentate systems, it would appear that the ratio k_{35}/k_{34} is of the order of magnitude of 1.

Unusually large experimental rate constants ($\sim 10^7$ M^{-1} s^{-1}), assigned originally to k_{43} but actually $k_{43} k_{35}/k_{34}$, have been observed for reactions of polyamines, such as ethylenediamine. To explain this, an internal conjugate base (ICB) mechanism was proposed[115] in which the free end of the amine forms a hydrogen bond to a coordinated water, giving it some hydroxide character, as shown in Eq. (3.84). This is supposed to make the remaining waters more labile to substitution according to the ICB proposal. Strangely, this rate enhancement is not observed with amino acids or with 2-methylaminopyridine.

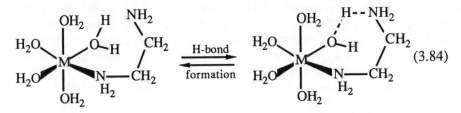

$$(3.84)$$

A reanalysis of these observations for the reaction of ethylenediamine with nickel(II)[116] has shown that $k_{43}k_{35}/k_{34} = 7.2 \times 10^6$ M^{-1} s^{-1}. One can estimate from the kinetics for the ethylamine reaction that $k_{43} \approx 900$ M^{-1} s^{-1} and $k_{34} \approx 15$ s^{-1}, so that $k_{35} \approx 1.2 \times 10^5$ s^{-1}. This value is about five times larger than the rate constant for water exchange on Ni(OH$_2$)$_6$$^{2+}$, but Hunt and co-workers.[117] found that the water exchange rate constant on (H$_2$O)$_5$NiNH$_3$$^{2+}$ is 2.5×10^5 s^{-1}. It is reasonable to expect that the water exchange on the monodentate ethylenediamine complex is similar to that of the NH$_3$ complex. Therefore, there is no anomalous reactivity for ethylenediamine relative to monodentate models.

The conclusion is that the high reactivity of the polyamines is due to a nonreacting ligand effect in which amines greatly increase the rate of water release from Ni(II). The work of Hunt et al. also established that a coordinated pyridine or carboxylate group does not have a strong labilizing influence on Ni(II). This explains the normal reactivity of the amino acids and pyridine amines.

Moore and co-workers[118] used stopped-flow nmr to measure the rate of first bond formation and chelate ring closing in the reactions of Al(DMSO)$_6$$^{3+}$ with 2,2'-bipyridine and 2,2':6',2"-terpyridine systems in nitromethane. The rate constant for first bond formation ($\sim 2 \times 10^3$ M^{-1} s^{-1}

at 25°C) is much larger than the DMSO exchange rate (5.3×10^{-2} s^{-1} at 25°C), presumably because of strong preassociation of the reactants. The chelate ring-closing rate constants ($\sim 10^{-2}$ s^{-1}) are smaller than the solvent exchange rate constant, possibly because of steric hindrance to twisting the second pyridine ring into a conformation suitable for chelation. In the same study, the reaction with 1,10-phenanthroline proceeded in a one-step process to the chelate. This may be because the fused ring of phenanthroline forces the two donor nitrogen atoms to be oriented for chelate formation.

3.9.b Kinetics and the Chelate Effect

It has been observed many times that the formation constant for a chelate such as $(H_2O)_4Ni(en)^{2+}$ ($K_f = 5 \times 10^7$ M^{-1}) is substantially greater than that for the monodentate analogue such as $(H_2O)_4Ni(NH_3)_2^{2+}$ ($K_f = \beta_2 = 1.5 \times 10^5$ M^{-2}). There have been several approaches to explaining this effect, such as differences in entropy change due to the different number of particles involved and/or a very large forward rate constant for ring closing because of local concentration effects.

It is instructive to consider the chelate formation process from a kinetic analysis of the individual steps given in Eq. (3.85):

$$ M + L{-}L \underset{k_{34}}{\overset{k_{43}}{\rightleftharpoons}} M{-}L{-}L \underset{k_{53}}{\overset{k_{35}}{\rightleftharpoons}} M\!\!\left\langle{\begin{array}{c}L\\ \\L\end{array}}\right. \qquad (3.85) $$

For the ethylenediamine system at 25°C, the analysis in the preceding section gave $k_{43} \approx 900$ M^{-1} s^{-1}, $k_{34} \approx 15$ s^{-1}, and $k_{35} \approx 1.2 \times 10^5$ s^{-1}, and from direct measurements of the dissociation kinetics, $k_{53} \approx 0.14$ s^{-1}. The small value of k_{53} for chelate ring opening is unexpected because, from a mechanistic standpoint, one would expect k_{53} to increase relative to k_{34} roughly in proportion to k_{35}/k_{43}. But k_{53} is much smaller than k_{34}. One reaches a similar conclusion for glycine, for which the rate constants at 25°C are given in Scheme 3.7.

Scheme 3.7

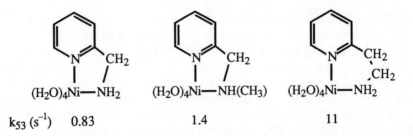

k_{53} (s^{-1}) 0.83   1.4   11

Figure 3.5. Rate constants for chelate ring opening in $(H_2O)_4Ni$(amino-pyridine)$^{2+}$.

It has been noted for some time that the magnitude of the chelate effect decreases as the chelate ring size increases. For example, the formation constants[119] for the aquanickel(II) complexes of ethylenediamine and trimethylenediamine (1,3 diaminopropane) are 2.5×10^7 and 2×10^6 M^{-1}, respectively. The six-membered chelate ring complex has about a 10 times smaller formation constant, despite the fact that trimethylenediamine is a stronger base toward the proton. This effect has traditionally been assigned to a smaller chelate ring-closing rate constant (k_{35}) for the longer chelate arm in the monodentate intermediate. However, the preceding analysis implies that the effect might lie in a larger rate constant for chelate ring opening (k_{53}) with the larger chelate ring.

Some evidence for this is found for amino-pyridine systems,[120,121] as shown in Figure 3.5. The rate constant for ring opening of the six membered chelate is about 10 times larger than that for either of the five-membered rings. A methyl group on the coordinated amine has a more modest effect that could be ascribed to steric acceleration of a dissociative process.

The conclusion is that chelate ring opening (k_{53}) *is an unexpectedly slow process,* and this accounts for the high formation constants of chelate complexes, since

$$K_f = \frac{k_{43}\, k_{35}}{k_{34}\, k_{53}} \tag{3.86}$$

Previous explanations have assumed that k_{35} is unusually large. It is somewhat ironic that k_{53} has received very little attention, since kinetic studies in this area have concentrated on entering group effects.

References

1. Langford, C. H.; Gray, H. B. *Ligand Substitution Processes*; Benjamin, Inc.: New York, 1966.
2. Eigen, M. Z. *Electrochem.* **1960**, *64*, 115.
3. Fuoss, R. M. *J. Am. Chem. Soc.* **1958**, *80*, 5059.

4. Haim, A.; Wilmarth, W. K. *Inorg. Chem.* **1962**, *1*, 573, 583.
5. Haim, A.; Grassi, R. J.; Wilmarth, W. K. *Adv. Chem. Ser.* **1965**, *49*, 31.
6. Burnett, M. G.; Gilfillian, W. M. *J. Chem. Soc., Dalton Trans.* **1981**, 1578.
7. Haim, A. *Inorg. Chem.* **1982**, *21*, 2887.
8. Abou-El-Wafa, M. H. M.; Burnett, M. G.; McCullagh, J. F. *J. Chem. Soc., Dalton Trans.* **1987**, 2311 and references therein.
9. Bradley, S. M.; Doine, H.; Krouse, H. R.; Sisley, M. J.; Swaddle, T. W. *Aust. J. Chem.* **1988**, *41*, 1323.
10. Abou-El-Wafa, M. H. M.; Burnett, M. G.; McCullagh, J. F. *J. Chem. Soc., Dalton Trans.* **1986**, 2083.
11. Robb, D.; Steyn, M. M. De V.; Kruger, H. *Inorg. Chim. Acta* **1969**, *3*, 383.
12. Stranks, D. R.; Yandell, J. K. *Inorg. Chem.* **1970**, *9*, 751.
13. Seibles, L.; Deutsch, E. D. *Inorg. Chem.* **1977**, *16*, 2273.
14. O'Brien, P.; Sweigart, D. A. *Inorg. Chem.* **1982**, *21*, 2094.
15. Sargeson, A. M.; Jordan, R. B. *Inorg. Chem.* **1965**, *4*, 431.
16. Buckingham, D. A.; Clark, C. R.; Lewis, T. W. *Inorg. Chem.* **1979**, *18*, 1985.
17. Jackson, W. G.; McGregor, B. C.; Jurisson, S. S. *Inorg. Chem.* **1987**, *26*, 1286; Jackson, W. G.; Hookey, C. N. *Inorg. Chem.* **1984**, *23*, 668, 2728.
18. Brasch, N. E.; Buckingham, D. A.; Clark, C. R.; Finnie, K. S. *Inorg. Chem.* **1989**, *28*, 4567.
19. Basolo, F.; Pearson, R. G. *Mechanisms of Inorganic Reactions*, 2nd ed.; Wiley & Sons: New York, 1967.
20. Jackson, W. G.; Dutton, B. H. *Inorg. Chem.* **1989**, *28*, 525.
21. House, D. A.; Powell, H. K. J. *Inorg. Chem.* **1971**, *10*, 1583.
22. Lawrance, G. A. *Inorg. Chem.* **1982**, *21*, 3687.
23. Curtis, N. J.; Lawrance, G. A.; van Eldik, R. *Inorg. Chem.* **1989**, *28*, 329.
24. Cattalini, L.; Ugo, R.; Orio, A. *J. Am. Chem. Soc.* **1968**, *90*, 4800.
25. Toma, H.; Malin, J. M. *J. Am. Chem. Soc.* **1972**, *94*, 4039.
26. Maresca, L.; Natile, G.; Calligaris, M.; Delise, P.; Randaccio, L. *J. Chem Soc., Dalton Trans.* **1976**, 2386.
27. Canovese, L.; Cattalini, L.; Uguagliati, P.; Tobe, M. L. *J. Chem. Soc., Dalton Trans.* **1990**, 867.
28. Pearson, R. G.; Sobel, H.; Songstad, J. *J. Am. Chem. Soc.* **1968**, *90*, 319.
29. Schwarzenbach, G.; Shellenberg, M. *Helv. Chim. Acta* **1965**, *48*, 28.
30. Edwards, J. O. *J. Am. Chem. Soc.* **1954**, *76*, 1540.
31. Mayer, U.; Gutmann, V. *Adv. Inorg. Chem. Radiochem.* **1975**, *17*, 189; Gutmann, V. *Electrochim. Acta* **1976**, *21*, 661.
32. Maria, P.-C.; Gal, J.-F. *J. Phys. Chem.* **1985**, *89*, 1296.
33. Kamlet, M. J.; Gal, J.-F.; Maria, P.-C.; Taft, R. W. *J Chem. Soc., Perkin Trans. 2* **1985**, 1583.
34. Abraham, M. H.; Grellier, P. L.; Prior, D. V.; Taft, R.W.; Morris, J. J.; Taylor, P. J.; Laurence, C.; Berthelot, M.; Doherty, R. M.; Kamlet, M. J.; Aboud, J.-L. M.; Sraidi, K.; Guihéneuf, G. *J. Am. Chem. Soc.* **1988**, *110*, 8534.
35. Catalán, J.; Gómez, J.; Couto, A.; Laynez, J. *J. Am. Chem. Soc.* **1990**, *112*, 1678.

36. Drago, R. S. *Coord. Chem. Rev.* **1980**, *33*, 251; Drago, R. S.; Wong, N.; Bilgrien, C.; Vogel, G. C. *Inorg. Chem.* **1987**, *26*, 9.
37. Lim, Y. Y.; Drago, R. S.; Babich, M. W.; Wong, N.; Doan, P. E. *J. Am. Chem. Soc.* **1987**, *109*, 169.
38. Pearson R. G. *J. Chem. Educ.* **1968**, *45*, 643.
39. Arhland, S.; Chatt, J.; Davies, N. R. *Quart. Rev. Chem. Soc.* **1958**, *12*, 265.
40. Pearson, R. G. *Inorg. Chem.* **1988**, *27*, 734.
41. Doan, P. E.; Drago, R. S. *J. Am. Chem. Soc.* **1984**, *106*, 2772.
42. Kamlet, M. J.; Gal, J.-F.; Maria, P.-C.; Taft, R. W. *J. Chem Soc., Perkin Trans. 2* **1985**, 1583.
43. Drago, R. S. *Inorg. Chem.* **1990**, *29*, 1379.
44. Maria, P.-C.; Gal, J.-F.; Francheschi, J.; Fargin, E. *J. Am. Chem. Soc.* **1987**, *109*, 483.
45. Langford, C. H. *Inorg. Chem.* **1965**, *4*, 265.
46. Haim, A. *Inorg. Chem.* **1970**, *9*, 426.
47. Swaddle, T. W.; Gustalla, G. *Inorg. Chem.* **1968**, *7*, 1915.
48. Tucker, M. A.; Colvin, C. B.; Martin, D. S., Jr. *Inorg. Chem.* **1964**, *3*, 1373.
49. Wax, M. J.; Bergman, R. G. *J. Am. Chem. Soc.* **1981**, *103*, 7028.
50. Rerek, M. E.; Basolo, F. *J. Am. Chem. Soc.* **1984**, *106*, 5908.
51. Parris, M.; Wallace, W. J. *Can. J. Chem.* **1969**, *7*, 2257.
52. Lay, P. A. *Inorg. Chem.* **1987**, *26*, 2144.
53. Basolo, F.; Chatt, J.; Gray, H. B.; Pearson, R. G.; Shaw, B. L. *J. Chem. Soc.* **1961**, 2207.
54. Tolman, C. A. *Chem. Rev.* **1977**, *77*, 313.
55. DeSanto, J. T.; Mosbo, J. A.; Storhoff, B. N.; Bock, P. L.; Bloss, R. E. *Inorg. Chem.* **1980**, *19*, 3086.
56. Rahman, Md. M.; Liu, H. Y.; Prock, A.; Giering, W. P. *Organometallics* **1987**, *6*, 650.
57. Swaddle, T. W.; Stranks, D. R. *J. Am. Chem. Soc.* **1972**, *94*, 8357.
58. Helm, L.; Elding, L. I.; Merbach, A. E. *Inorg. Chem.* **1985**, *24*, 1719.
59. van Eldik, R.; Asano, T.; Le Noble, W. J. *Chem. Rev.* **1989**, *89*, 549.
60. Twigg, M. V. *Inorg. Chim. Acta* **1977**, *24*, L84.
61. Garrick, F. J. *Nature* **1937**, *139*, 507.
62. Tobe, M. L. *Acc. Chem. Res.* **1970**, *3*, 377; *Adv. Inorg. Bioinorg. Mech.* **1983**, *2*, 1.
63. Sargeson, A. M. *Pure Appl. Chem.* **1973**, *33*, 527.
64. Buckingham, D. A.; Clark, C. R.; Lewis, T. W. *Inorg. Chem.* **1979**, *18*, 2041.
65. Appleton, T. G.; Clark, H. C.; Manzer, L. E. *Coord. Chem. Rev.* **1973**, *10*, 335.
66. Coyle, B. A.; Ibers, J. A. *Inorg. Chem.* **1972**, *11*, 1105.
67. Chatt, J.; Duncanson, L.; Shaw, B.; Venanzi, L. *Disc. Faraday Soc.* **1958**, *26*, 131; Adams, D. M.; Chatt, J.; Shaw, B. *J. Chem. Soc.* **1960**, 2047.
68. Parshall, G. W. *J. Am. Chem. Soc.* **1966**, *88*, 704.
69. Zumdahl, S. S.; Drago, R. S. *J. Am. Chem. Soc.* **1970**, *90*, 6669.
70. Armstrong, D. R.; Fortune, R.; Perkins, P. G. *Inorg. Chim. Acta* **1974**, *9*, 9.

71. Davy, R. D.; Hall, M. B. *Inorg. Chem.* **1989**, *28*, 3524.
72. Hunt, J. P.; Rutenberg, A. C.; Taube, H. *J. Am. Chem. Soc.* **1952**, *74*, 268.
73. Posey, F. A.; Taube, H. *J. Am. Chem. Soc.* **1953**, *75*, 4099.
74. Palmer, D. A.; van Eldik, R. *Chem. Rev.* **1983**, *83*, 561.
75. Van Eldik, R.; Harris, G. M. *Inorg. Chem.* **1980**, *19*, 880; Dasgupta, T. P.; Harris, G. M. *Inorg. Chem.* **1984**, *23*, 4398.
76. Joshi, V. K.; van Eldik, R.; Harris, G. M. *Inorg. Chem.* **1986**, *25*, 2229.
77. Murmann, R. K.; Taube, H. *J. Am. Chem. Soc.* **1956**, *78*, 4886.
78. Jackson, W. G.; Lawrance, G. A.; Lay, P. A.; Sargeson, A. M. *Inorg. Chem.* **1980**, *19*, 904.
79. Andrade, C.; Taube, H. *J. Am. Chem. Soc.* **1964**, *86*, 1328.
80. Buckingham, D. A.; Foster, D. M.; Sargeson, A. M. *J. Am. Chem. Soc.* **1969**, *91*, 4102.
81. Buckingham, D. A.; Keene, F. R.; Sargeson, A. M. *J. Am. Chem. Soc.* **1974**, *96*, 4981.
82. Buckingham, D. A. In *Biological Aspects of Inorganic Chemistry* ; Addison, A. W.; Cullen, W. R.; Dolphin, D.; James, B. R., Eds.; Wiley Interscience: New York, 1977; pp. 141–196.
83. Boreham, C. J.; Buckingham, D. A.; Francis, D. J.; Sargeson, A. M.; Warner, L. G. *J. Am. Chem. Soc.* **1981**, *103*, 1975.
84. Pinnel, D.; Wright, G. B.; Jordan, R. B. *J. Am. Chem. Soc.* **1972**, *94*, 6104; Buckingham, D. A.; Keene, F. R.; Sargeson, A. M. *J. Am. Chem. Soc.* **1973**, *95*, 5649.
85. Dixon, N. E.; Fairlie, D. P.; Jackson, W. G.; Sargeson, A. M. *Inorg. Chem.* **1983**, *22*, 4038.
86. Ellis, W. R., Jr.; Purcell, W. L. *Inorg. Chem.* **1982**, *21*, 834.
87. Anderson, B.; Milburn, R. M.; MacB. Harrowfield, J.; Robertson, G. B.; Sargeson, A. M. *J. Am. Chem. Soc.* **1977**, *98*, 2652.
88. Clark, C. R.; Tasker, R. F.; Buckingham, D. A.; Knighton, D. R.; Harding, D. R. K.; Hancock, W. S. *J. Am. Chem. Soc.* **1981**, *103*, 7023.
89. Sutton, P. A.; Buckingham, D. A. *Acc. Chem. Res.* **1987**, *20*, 357.
90. Plane, R. A.; Hunt, J. P. *J. Am. Chem. Soc.* **1954**, *76*, 5960; Ibid. **1957**, *79*, 3343.
91. Swift, T. J.; Connick, R. E. *J. Chem. Phys.* **1962**, *37*, 307.
92. Merbach, A. E. *Pure Appl. Chem.* **1982**, *54*, 1479; Ibid. **1987**, *59*, 161.
93. Taube, H. *Chem. Rev.* **1952**, *50*, 69.
94. Basolo, F.; Pearson, R. G. *Mechanisms of Inorganic Reactions*, 2nd ed.; Wiley & Sons: New York, 1967; pp. 65–80.
95. Breitschwerdt, K. *Ber. Bunsenges. Phys. Chem.* **1968**, *72*, 1046.
96. Spees, S. T. Jr.; Perumareddi, J. R.; Adamson, A. W. *J. Am. Chem. Soc.* **1968**, *90*, 6626.
97. Burdett, J. K. *Adv. Inorg. Chem. Radiochem.* **1978**, *21*, 113.
98. Mønsted, O. *Acta Chem. Scand.* **1978**, *A32*, 297.
99. Rode, B. M.; Reihnegger, G. J.; Fujiwara, S. *J. Chem. Soc., Faraday Trans.* 2 **1980**, *76*, 1268.
100. Connick, R. E.; Alder, B. J. *J. Phys. Chem.* **1983**, *87*, 2764.
101. Swaddle, T. W. *Inorg. Chem.* **1983**, *22*, 2663.
102. Rusnak, L. L.; Yang, E. S.; Jordan, R. B. *Inorg. Chem.* **1978**, *17*, 1810.

103. Eigen, M.; Wilkins, R. G. *Adv. Chem. Ser.* **1965**, *49*, 55; Kustin, K.;
     Swinehart, J. *Prog. Inorg. Chem.* **1970**, *13*, 107; Hewkin, D. J.; Prince,
     R. H. *Coord. Chem. Rev.* **1970**, *5*, 45; Swaddle, T. W. *Adv. Inorg.
     Bioinorg. Mech.* **1983**, *2*, 96; Hoffmann, H. *Pure Appl. Chem.* **1975**, *41*,
     327; Coetzee, J. F. In *Solute–Solvent Interactions*, Coetzee, J. F.; Ritchie,
     C. D., Eds.; Marcel Dekker: New York, 1976; Vol. 2, pp. 331–418.
104. Wilkins, R. G. *Adv. Inorg. Bioinorg. Mech.* **1983**, *2*, 139; Eigen, M. *Pure
     Appl. Chem.* **1963**, *6*, 7.
105. Wilkins, R. G. *Acc. Chem. Res.* **1970**, *3*, 407.
106. Wilkins, R. G.; Eigen, M. *Adv. Chem. Ser.* **1965**, *49*, 55.
107. Meyer, F. K.; Newman, K. E.; Merbach, A. E. *J. Am. Chem. Soc.* **1979**,
     101, 5588; Ducommun, Y.; Newman, K. E.; Merbach, A. E. *J. Am. Chem.
     Soc.* **1980**, *19*, 3696.
108. Chaudhuri, P.; Diebler, H. *J. Chem Soc., Dalton Trans.* **1986**, 1693 and
     references therein.
109. Hugi, A. D.; Helm, L.; Merbach, A. E. *Inorg. Chem.* **1987**, *26*, 1763.
110. Patel, R. C.; Diebler, H. *Ber. Bunsenges. Phys. Chem.* **1972**, *76*, 1035.
111. Fogg, P. G. T.; Hall, R. J. *J. Chem. Soc. A* **1971**, 1365.
112. Monzyk, B.; Crumblis, A. L. *J. Am. Chem. Soc.* **1979**, *101*, 6203.
113. Letter, J. E., Jr.; Jordan, R. B. *J. Am. Chem. Soc.* **1975**, *97*, 2381.
114. Cassatt, J. C.; Johnson, W. A.; Smith, L. M.; Wilkins, R. G. *J. Am.
     Chem. Soc.* **1972**, *94*, 8399.
115. Rorabacher, D. B. *Inorg. Chem.* **1966**, *5*, 1891.
116. Jordan, R. B. *Inorg. Chem.* **1976**, *15*, 748.
117. Desai, A. G.; Dodgen, H. W.; Hunt, J. P. *J. Am. Chem. Soc.* **1969**, *91*,
     5001; Ibid. **1970**, *92*, 798.
118. Brown, A. J.; Howarth, O. W.; Moore, P.; Parr, W. J. E. *J. Chem. Soc.,
     Dalton Trans.* **1978**, 1776.
119. Martell, A. E.; Smith, R. M. *Critical Stability Constants*; Plenum: New
     York, 1977; Vol. 2.
120. Hubbard, C. D.; Palaitis, W. *J. Coord. Chem.* **1979**, *9*, 107.
121. Jordan, R. B. *J. Coord. Chem.* **1980**, *10*, 239.

4

Stereochemical Change

The kinetic and mechanistic aspects of this general area tend to be strongly dependent on the particular system. This makes general treatments and explanations impossible, at least at the current stage of understanding. Various aspects of this area have been summarized in some general reviews.[1-5]

4.1 TYPES OF LIGAND REARRANGEMENTS

Ligands bonded to a metal can undergo a number of structural changes that do not involve complete breaking of the metal–ligand bond(s). Such processes are the subject of the following sections.

4.1.a Conformational Change

Many chelate ligands have conformers that can interconvert. For example, the conformers of ethylenediamine interchange by rotation about the carbon–carbon bond:

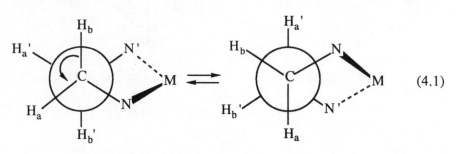

$$(4.1)$$

The H_a and H_a' protons are magnetically different from the H_b and H_b' protons, so their interconversion can be studied by nmr. In simple systems this interconversion seems to be quite rapid ($k > 10^6$ s^{-1}). However, if there is some constraint (e.g., CH$_3$ groups) or if the coordinating atoms are part of a larger chelate system, then interconversion is slow enough to be detected by nmr.[6]

4.1.b   Coordination Geometry Change

There are several nickel(II) complexes, such as those with the ligands shown below, which exist as equilibrium mixtures of paramagnetic *tetrahedral* and diamagnetic *square planar* isomers.

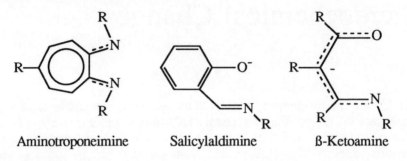

Aminotroponeimine   Salicylaldimine   ß-Ketoamine

The planar–tetrahedral interconversions are rapid (k ~ 10^4–10^6 s^{-1}, $\Delta H^* \approx$ 10 kcal mol^{-1}, $\Delta S^* \approx 0$). The mechanism for the transformation is thought to be intramolecular rearrangement with no metal–ligand bond breaking.[7]

Octahedral to *square planar* transformations have also been observed. Again nickel(II) complexes provide the most examples, with systems having a four-coordinate ligand in the square plane and two solvent molecules occupying the other octahedral positions, as shown in Eq. (4.2):

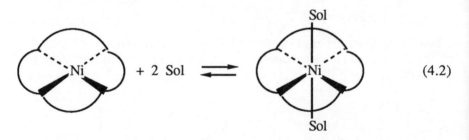

$$+\ 2\ Sol \rightleftharpoons \tag{4.2}$$

These interconversions are quite rapid and have been studied by laser T-jump.[8]

A *tetrahedral* to *octahedral* conversion has been studied for the cobalt(II)[9] complex in Eq. (4.3):

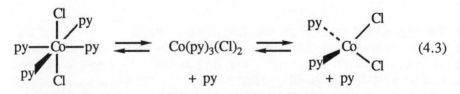

$$\rightleftharpoons\quad Co(py)_3(Cl)_2 \rightleftharpoons \tag{4.3}$$

$$+\ py \qquad\qquad +\ py$$

The results have been interpreted in terms of a five-coordinate intermediate and the reactions are rapid, as expected for the labile cobalt(II) (d^7) ion.

4.1.c Linkage Isomerism

Linkage isomerism has been studied with inert metal ions and is possible, in principle, for any metal with a ligand that has more than one type of atom with an unshared electron pair. The following equations give some common examples:[10,11]

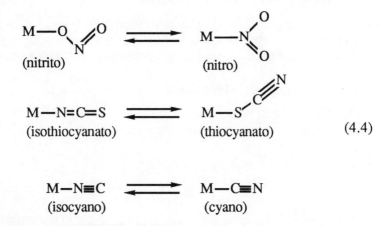

(nitrito)   (nitro)

(isothiocyanato)   (thiocyanato)   (4.4)

(isocyano)   (cyano)

With inert octahedral complexes these reactions appear to be intramolecular. This can be demonstrated by allowing the reaction to proceed in the presence of uncomplexed isotopically labeled ligand. For example, the rearrangement of $(NH_3)_5CoSCN^{2+}$ in water in the presence of $^{15}NCS^-$ occurs without the incorporation of $^{15}NCS^-$ in the product $(NH_3)_5CoNCS^{2+}$. It is not certain whether the rearrangement proceeds by metal–ligand bond breaking followed by ligand rotation and bond making to the new atom or through slippage,[12] as shown in Eq. (4.5):

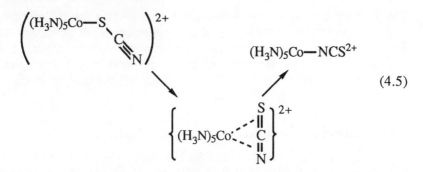

(4.5)

An intermolecular rearrangement has been established for $(Et_4dien)PdSCN^+$, with the reaction probably proceeding via a solvent intermediate that reanates to give the isomerized product.[13]

Data on the linkage isomerization of $(NH_3)_5Co(ONO)^{2+}$ in water were given in Chapter 1. The reaction rate has been shown[14] to have a significant solvent dependence, which has been interpreted in terms of the Lewis basicity, acidity, and polarity of the solvents.

4.2 OPTICAL AND GEOMETRICAL ISOMER INTERCONVERSION

The interconversion of geometrical isomers (cis, trans, fac, mer) and the racemization of optical isomers ($\lambda - \Delta$) can proceed by two general mechanisms, *ligand dissociation* or *intramolecular rearrangement*.

4.2.a Dissociation Mechanisms

If the complex is an octahedral bis chelate system, $(AA)_2M(X)_2$, then two trigonal bipyramidal intermediates are possible, as shown in Scheme 4.1.

Scheme 4.1

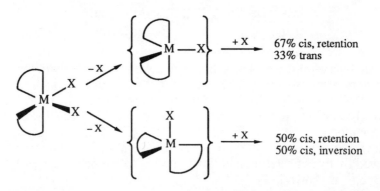

The product distributions in Scheme 4.1 are based on the assumption of a statistical attack along the edges of the trigonal plane of the intermediate. Since the lower intermediate has a plane of symmetry, it must lead to racemization. Similar intermediates can be drawn for tris chelate systems if one end of a chelate dissociates. This is called *one-ended dissociation* and is actually an intramolecular process.

4.2.b Intramolecular Rearrangement Mechanisms

This mechanism will be illustrated for tris chelates of the type $M(AA)_3$ but can be easily extended to other systems. The two most commonly considered rearrangements involve rotating one trigonal face of the octahedron by 120° relative to the opposite trigonal faces. The *Bailar* or *trigonal twist* involves rotation about a C_3 axis, as shown in Figure 4.1, and the *rhombic* or *Ray–Dutt bend* is rotation about an imaginary C_3 axis, as shown in Figure 4.2.

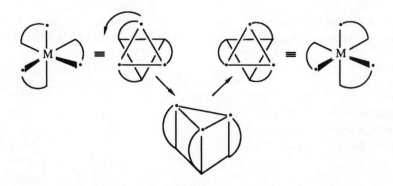

Figure 4.1. The Bailar or trigonal twist rearrangement for an M(AA)$_3$ complex.

The differentiation of the dissociation and intramolecular mechanisms is usually based on a comparison of the rates of ligand dissociation (exchange or solvolysis) and racemization. If the rates are quite similar, dissociation is assumed, whereas if racemization is much faster, then an intramolecular mechanism is operating.

For the tris(oxalato)Rh(III) ion in water, Damrauer and Milburn[15] have studied the kinetics of racemization, aquation, and $^{18}OH_2$ exchange of the oxalate oxygens. The racemization and aquation have acid-dependent pseudo-first-order rate constants given by

$$k_{exp} = k_2 [H^+] + k_3 [H^+]^2 \qquad (4.6)$$

The oxygen exchange is first order in [H$^+$] and the inner, coordinated oxygens are exchanged much more slowly than the outer oxygens. The kinetic results are summarized in Table 4.1.

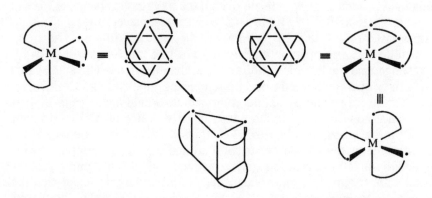

Figure 4.2. The rhombic or Ray–Dutt bend rearrangement for an M(AA)$_3$ complex.

Table 4.1. Rate Constants (50°C) and Activation Parameters for the Oxygen Exchange, Racemization, and Aquation of $Rh(C_2O_4)_3^{3-}$ in 0.54 M $NaClO_4/HClO_4$

Reaction	k_2 $(M^{-1} s^{-1})$	k_3 $(M^{-2} s^{-1})$	ΔH^* (kcal mol^{-1})	ΔS^* (cal mol^{-1} K^{-1})
Outer O exchange	1.1×10^{-3}		16.9	−20.0
Inner O exchange[a]	3.6×10^{-5}		23.5	−6.4
Racemization[a]	2.0×10^{-5}		23.0	−9.1
		1.1×10^{-4}	27.9	+9.6
Aquation[a]	5.6×10^{-7}		25.6	−8.1
		5.9×10^{-6}	25.7	−3.1

[a] Values recalculated from the data in Damrauer, L.; Milburn, R. M. *J. Am. Chem. Soc.* **1971**, *93*, 6481.

Damrauer and Milburn noted the similarity in the kinetic parameters of k_2 for inner oxygen exchange and racemization and suggested that both processes are proceeding through a common intermediate formed by acid-catalyzed one-ended dissociation of oxalate. For $K_3[Rh(C_2O_4)_3] \cdot 4.5H_2O$, the crystal structure[16] reveals that the O—Rh—O angles are about 83°. This makes it seem more probable that racemization and inner oxygen exchange proceed by a square pyramidal intermediate rather than by a symmetrical trigonal pyramidal one (Scheme 4.1), which requires one O—Rh—O angle of 120°. It is noteworthy that the kinetics of the outer oxygen exchange are similar to those of oxalic acid,[17] for which $\Delta H^* = 15.1$ kcal mol^{-1}, $\Delta S^* = -22.7$ cal mol^{-1} K^{-1}, and k (50°C) = 4×10^{-3} s^{-1}. Palmer and Kelm[18] found $\Delta V^* = -6.3$ cm^3 mol^{-1} for the aquation of $Rh(C_2O_4)_3^{3-}$ in 1.0 M H$^+$ where k_3 dominates.

Tris(oxalato)chromium(III) is much more reactive than its rhodium(III) analogue. Odell et al.[19] indicate that all the oxygens exchange at the same rate (k = 1.26×10^{-3} s^{-1}, 1.0 M HClO$_4$, 25°C) and racemization is about 10 times faster (k = 1.38×10^{-2} s^{-1}) under the same conditions. The authors conclude that ring opening and closing must be much faster than oxygen exchange, and the latter is 2.4 times faster than for free oxalic acid at 25°C. Structures of chromium(III)–oxalate complexes[20] indicate O—Cr—O angles of ~83°, and a square pyramidal intermediate would again seem probable. Lawrance and Stranks[21] found a very negative value of $\Delta V^* = -16.3$ cm^3 mol^{-1} for racemization of Cr(ox)$_3^{3-}$ in 0.05 M HCl. They rationalize this as mainly due to solvent electrostriction around the -CO$_2^-$ group in the one-ended dissociation transition state. Lawrance and Stranks imply that their results are for an [H$^+$]-independent reaction, but the [H$^+$]-catalyzed path dominates the kinetics below pH 3. For Cr(phen)$_2$(ox)$^+$ and Cr(bpy)$_2$(ox)$^+$, the values of ΔV^* are −1.5 and −1.0 cm^3 mol^{-1}, respectively, and an intramolecular twist is proposed by Lawrance and Stranks.

Table 4.2. Activation Parameters for the Racemization and Aquation of Some Complexes of Tris(o-phenanthroline)

Complex	Reaction	ΔH^* (kJ mol^{-1})	ΔS^* (J mol^{-1} K^{-1})	ΔV^* (cm^3 mol^{-1})
Fe(phen)$_3{}^{2+}$	Racemization	118 ($\pm$ 3)	89 ($\pm$ 8)	15.6 ($\pm$ 0.3)
	Aquation	135 ($\pm$ 2)	117 ($\pm$ 8)	15.4 ($\pm$ 0.3)
Ni(phen)$_3{}^{2+}$	Racemization	105 ($\pm$ 1)	12 ($\pm$ 3)	−1.5 ($\pm$ 0.3)
	Aquation	102 ($\pm$ 2)	3 ($\pm$ 6)	−1.2 ($\pm$ 0.2)
Cr(phen)$_3{}^{3+}$	Racemization	94 ($\pm$ 4)	−56 ($\pm$ 3)	3.3 ($\pm$ 0.3)

The racemization of tris(N,N'-dimethylethylenediamine)Ni(II) appears to proceed by an intramolecular rotation because bond rupture should produce some meso isomer, which is not observed. The kinetic similarity of this and the Ni(en)$_3{}^{2+}$ racemization[22] suggests that the latter also is intramolecular. On the other hand, the tris(bipyridyl) and tris(o-phenanthroline) complexes of nickel(II)[23] and iron(II)[24] have very similar rates of racemization, ligand exchange, and aquation, and therefore are assumed to proceed by dissociation. It seems surprising that these more rigid chelates would choose dissociation while the more flexible aliphatic diamines go by intramolecular rotation. It should be remembered that the aliphatic amines are much stronger bases than the aromatic amines, and thus the former may form stronger bonds to the metal, making dissociation less favorable.

Comparisons of volumes of activation in Table 4.2 from the work of Lawrance and Stranks[25] provide some further insights. The authors suggest that the small values of ΔV^* for Ni(II) and Cr(III) imply an intramolecular twist mechanism. To rationalize the large ΔV^* for Fe(II), they suggest that there is a spin state change accompanying the intramolecular twist (i.e., the system goes from low-spin d^6 in the ground state to high-spin d^6 in the transition state). This causes the metal-to-ligand bond lengths to increase.

Such spin equilibria have been studied for the iron(II) complexes of the ligands shown below. For (papth)$_2$Fe(II), $\Delta V^o = 11$ cm^3 mol^{-1},[26] and the kinetics for the low- to high-spin direction gave k = 1.7 x 10^7 s^{-1} (25°C),

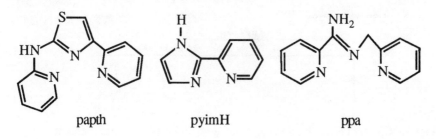

papth   pyimH   ppa

$\Delta H^* = 31.7$ kJ mol^{-1}, and $\Delta S^* = 0.37$ J mol^{-1} K^{-1}. Similar kinetic parameters have been found[27] for the pyimH and ppa iron(II) chelates, which have ΔV^* values in the 2 to 9 cm^3 mol^{-1} range with a significant solvent dependence.

There have been several attempts to provide a theoretical framework for describing geometrical isomerization and racemization. Vanquickenborne and Pierloot[28] used ligand field theory to calculate the electronic energies of the intermediates proposed in the dissociative and Bailar twist mechanisms for low-spin d^6 systems.

The d orbital energy diagrams from the analysis of Vanquickenborne and Pierloot for the Bailar twist are shown in Figure 4.3. The values of σ and π are measures of the σ and π donor strength of the ligand (e.g., for en, $\sigma = 94.3$ kJ mol^{-1}, $\pi = 0$; for H$_2$O, $\sigma = 78.5$ kJ mol^{-1}, $\pi = 12.85$ kJ mol^{-1}). The prediction is that these systems should go through a spin state change to give a more stable transition state. To calculate the ΔH^*, one must take into account electron repulsion–delocalization factors through the Racah parameters (B and C) in ligand field theory. The diagram gives a change of $6\sigma - 8\pi$ for formation of the transition state, and $\Delta H^* = 6\sigma - 8\pi - 5B - 8C$ for a d^6 system. For Co(III), B = 7.14 kJ mol^{-1}, C = 44 kJ mol^{-1}, and $\Delta H^* \approx 180$ kJ mol^{-1} for Co(en)$_3^{3+}$. Oxygen donor ligand systems, with smaller σ values and π values >0, should twist more readily than nitrogen donor systems.

For the dissociative mechanism, it was assumed that a square-based pyramid (C_{4v}) would form first and would rearrange to a trigonal bipyramid (D_{3h}) if some stereochemical change is observed. The orbital energies are given in Figure 4.4. The theory again predicts a spin state change in the D_{3h} intermediate. As before, the diagram gives a change of $5\sigma - 6.5\pi$ for the high-spin transition state, and $\Delta H^* = 5\sigma - 6.5\pi - 5B - 8C \approx 85$ kJ mol^{-1} for the Co(en)$_3^{3+}$. A comparison of the ΔH^* values from the two mechanisms predicts that the rearrangement should go by a dissociative mechanism through a high-spin transition state.

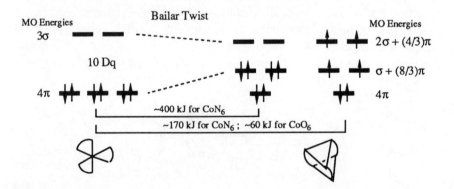

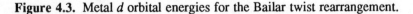

Figure 4.3. Metal d orbital energies for the Bailar twist rearrangement.

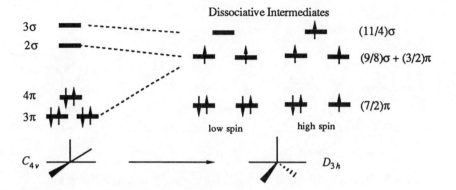

Figure 4.4. Metal *d* orbital energies for the dissociative rearrangement mechanism.

The theory has been extended to Co(en)$_2$(A)(X) systems,[29] but the results are more complex because of the different ligand types. Vanquickenborne concludes for these systems that the controlling factor is $\sim(\sigma_A - 2\pi_A)$, so that better π-donor A ligands will reduce the C_{4v} to D_{3h} energy barrier and therefore give more stereochemical change in a dissociative mechanism. This prediction is consistent with observations that A ligands such as CN$^-$ or NH$_3$ give hydrolysis with retention, whereas if A is Cl$^-$ or OH$^-$, there is a considerable (~30 percent) change in configuration. The theory of Vanquickenborne predicts that the amount of rearrangement should increase with increasing temperature, but this aspect has not been tested experimentally.

For those interested in studying racemization, a major problem has always been the need to resolve the optical isomers. However, it is possible to avoid this problem by using a chelate with a chiral center and studying the system with nmr. An example is diisobutyrylmethanide, shown in Figure 4.5, where the isopropyl carbon is chiral and the view at the right down the C—C bond shows that the two CH$_3$ groups can never be magnetically equivalent. When inversion occurs in a tris chelate of this type, the two CH$_3$ groups are interchanged, and this is observed as a merging of the two nmr signals as the interconversion becomes fast on the nmr time scale; actually H–H coupling causes four peaks to collapse to two. The process is pictured

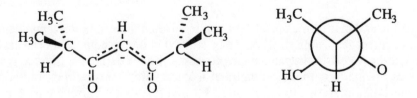

Figure 4.5. The diisobutyrylmethanide group with a chiral isopropyl carbon.

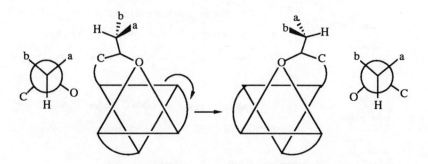

Figure 4.6. Interconversion of methyl groups a and b during a Bailar twist.

in Figure 4.6 for just one isopropyl group with the CH_3 groups designated as a and b. The same technique can be used for the H atoms in a $-CH_2CH_3$ chelate ring substituent.

4.2.c Rearrangements in Unsymmetrical Chelates

There are a large number of potential unsymmetrical chelates such as the following examples:

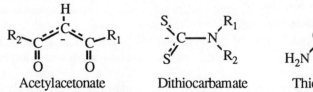

Acetylacetonate   Dithiocarbamate   Thioethanolamine

These have the added feature of having both the geometrical isomers, shown below, and optical isomers.

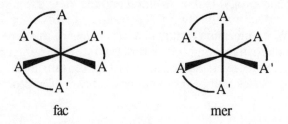

fac   mer

The fac and mer isomers each have optical isomers; thus, both racemization and geometrical isomerization can be studied in one system. This can serve to eliminate certain rearrangement processes. For example, the trigonal twist gives inversion without geometrical isomerization, but the rhombic bend gives both, as shown in Figure 4.7. Further results for various reasonable

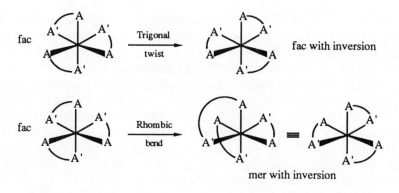

Figure 4.7. The products of the trigonal twist and rhombic bend for an unsymmetrical tris chelate.

D transition states are given by Wilkins,[30] as in the following examples:

Axial trigonal bipyramid → Isomerization + Inversion
Equatorial trigonal bipyramid → Isomerization + Inversion
Square-based pyramid → Some isomerization + Some inversion

Instead of this mechanistic approach, a pure permutational analysis can be used. The possibilities are given by Holm[31] and in more detail by Musher[32] and Eaton et al.[33] A detailed nmr analysis often leads to inconclusive mechanistic results. The Bailar twist is the most restrictive in terms of the rearrangement results. This mechanism has been confirmed for the complexes of Al(III), Ga(III), and Co(III), shown in Figure 4.8. In these systems, the Bailar (trigonal) twist may be favored by ground-state distortions toward the trigonal transition state because of the small bite distance (2.5 Å) of the ligand.[34] This distortion is measured by the twist angle (Ø in Figure 4.8) which is 60° for an octahedron and 0° for the trigonal transition state.

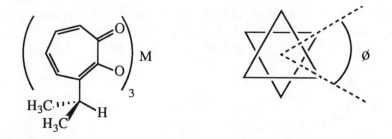

Figure 4.8. Complexes of Al(III), Ga(III), and Co(III),which rearrange by a Bailar twist and have a twist angle Ø.

Several iron(IV), iron(III), and ruthenium(III) dithiocarbamates also appear to use the trigonal twist mechanism.[35] It is fortunate that these systems rearrange by the one process that can be uniquely established by nmr because it gives inversion without fac–mer rearrangement.

The structural features that may favor the Bailar twist relative to the Ray–Dutt bend have been discussed by Rodger and Johnson.[36] Their approach considers the ligand bite distance (b) and the hard-sphere contact distance (l) of the coordinating atoms, and they conclude that the Bailar twist is favored if b is much smaller than l. Relevant structural information on such systems has been reviewed by Keppert.[37] This approach does not take into account any differences in strain in the chelate backbone.

4.3 STEREOCHEMICAL CHANGE IN FIVE-COORDINATE SYSTEMS

Most of the effort in the area of five-coordinate systems has been on trigonal bipyramid systems such as PF_5 and $Fe(CO)_5$. The interconversion is between the axial (a) and equatorial (e) positions in the following structure:

$$
\begin{array}{c}
\text{a} \\
\text{e} \overset{\displaystyle |}{\underset{\displaystyle |}{\overline{}}}\!\!\!\!\diagdown\!\!\begin{array}{l} e \\ e \end{array} \\
\text{a}
\end{array}
$$

The axial–equatorial conversion is generally quite rapid for ML_5 systems, but a series of $M(P(OR)_3)_5$ complexes have rates that are accessible on the nmr time scale. Meakin and Jesson[38] found values for ΔH^* of 8 to 12 kcal mol^{-1} and showed that the process required the simultaneous interchange of two axial and two equatorial substituents. The *pseudorotation mechanism* suggested by Berry,[39] and shown in Scheme 4.2, is consistent with the observations. The initially equatorial e' and e" ligands become axial, whereas the initially axial a and a' become equatorial. The product appears to be rotated by 90° relative to the starting structure. Note that the intermediate structure is close to a square-based pyramid.

Scheme 4.2

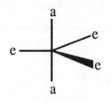

Ugi et al.[40] have suggested what is called the *turnstyle mechanism*, which is described in reaction 4.7. Since this process is also consistent with the observations, the two mechanisms cannot be distinguished experimentally. Theoretical arguments have favored the pseudorotation mechanism as providing the lower energy pathway for the interconversion.[41]

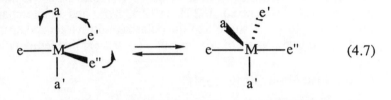

$$(4.7)$$

In the case of some transition-metal hydrides of the type HMP_4, the ground-state structure is a distorted trigonal bipyramid; it can be pictured as a tetrahedral arrangement of the P ligands around M with the H atom on one of the tetrahedral faces. A *tetrahedral jump mechanism* has been proposed[42] for such hydrides in which the H atom moves from one tetrahedral face to an edge and then to another face, as shown in Scheme 4.3. This mechanism is consistent with the nmr observations. If the H and the opposite P are considered as axial, then this process converts the axial P substituent to an equatorial position for each occurrence, and it is not the permutational equivalent of the pseudorotation process.

Scheme 4.3

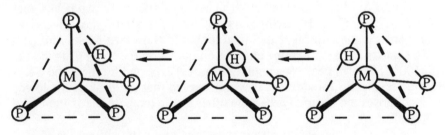

4.4 FLUXIONAL ORGANOMETALLIC COMPOUNDS

The mechanisms of geometrical and optical isomerization already discussed also apply to organometallic compounds, but these show some additional unique rearrangement processes. This area has been reviewed recently by Mann.[43,44]

4.4.a Iron Pentacarbonyl

Iron pentacarbonyl is a trigonal bipyramidal ML_5 complex for which the rearrangement mechanism has already been discussed. $Fe(CO)_5$ deserves special mention because of the amount of attention it has received. Sofar, the equatorial–axial exchange has proved to be too fast to measure by ^{13}C nmr, even down to −120°C.[45,46] The one condition on this is the assumption that there is a significant chemical shift difference between the axial and equatorial ^{13}C nuclei; this has been questioned by Mahnke et al.[47] A solid-state nmr study[48] down to 100 K indicated some fluxional behavior with an energy barrier of ~1 kcal. More recent solid state ^{13}C nmr results[49] indicate a shift difference of 182 Hz at 22.53 MHz and an interchange rate of <10^2 s^{-1} at −38°C. A study[50] of the polarized ultraviolet spectrum in a CO matrix at 20 K shows that interconversion is slow at that temperature.

4.4.b Fluxional Ring Systems

The term *ring whizzers* was coined by Cotton to describe the phenomenon that was first studied quantitatively by Bennett et al.[51] in the compound $Fe(CO)_2(\eta^5\text{-}C_5H_5)(\eta^1\text{-}C_5H_5)$, whose structure follows:

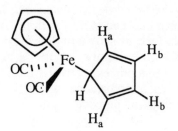

The ^1H nmr for this compound at 30°C in CS_2 shows only two peaks, one typical of the $\eta^5\text{-}C_5H_5$ hydrogens and one for all of the $\eta^1\text{-}C_5H_5$ hydrogens. If the latter is static, then it should show two peaks (due to a and b protons in the structure) plus the unique H on the same C as the Fe. At −100°C, the two peaks (+ coupling features) expected for the a and b sets become resolved. In addition, it was observed that the peak at the lower field collapses more rapidly than the other as the temperature is increased from −100°C.

These observations are consistent with the Fe moving around the $C_5H_5^-$ ring or the ring whizzing around the Fe, depending on your point of view. The mechanism could be a series of either 1,2 or 1,3 shifts of the Fe about the $\eta^1\text{-}C_5H_5$. If the shift were a random process, then all the peaks should collapse at the same rate. The two shift mechanisms are shown in Scheme 4.4, which starts at the center and goes through a 1,3 shift to the left and a 1,2 shift to the right. The change in magnetic type of the original H atoms is shown in Scheme 4.4 for each type of shift. For the 1,2 shift, the B' becomes a magnetically equivalent B, whereas for the 1,3 shift, the A becomes a magnetically equivalent A'. Therefore, the mechanism would be a 1,2 shift if the B resonance is collapsing more slowly and a 1,3 shift if the A resonance is the one collapsing more slowly.

Scheme 4.4

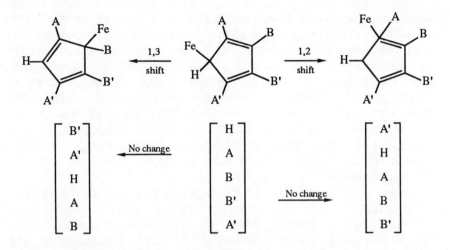

By analogy to other systems, Bennett et al. assigned the more slowly collapsing resonance to the B,B' hydrogens and concluded that the mechanism involved a 1,2 shift. However, the spectral assignment remained somewhat uncertain, and it is a common feature in this type of work that *the mechanism will rely directly on the validity of the spectral assignments for the low temperature, nonexchanging state of the system.*

To test the preceding assignment of the spectrum and mechanism, Cotton et al.[52] prepared the indenyl derivative in Eq. (4.8). They anticipated that a 1,2 shift would be unfavorable in this system because of the loss of aromatic resonance energy in the product.

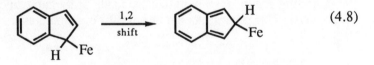

$$(4.8)$$

The $Fe(CO)_2(\eta^5\text{-}C_5H_5)(\eta^1\text{-indenyl})$ shows no fluxional behavior, and all the expected hydrogen resonances are observed in the room-temperature nmr spectrum.

4.4.c Symmetry Rules for Sigmatropic Shifts

As a result of the above and other work through the 1970s, it appeared that these fluxional processes prefer the 1,2 shift mechanism; this led to what has been termed "the principle of least motion" as being a major factor in determining the shift mechanism. Recent work has examined these processes in more detail and emphasized the orbital symmetry involved, following the Woodward–Hoffmann rules[53] for sigmatropic shifts in organic chemistry. The adaptation of these rules to inorganic transition-metal systems has had some success, but the inorganic systems have several potential complications that weaken the predictive power of the rules. The main problem is that *p* and *d* orbitals may be involved, and their symmetry gives predictions opposite to those derived from organic systems where only sigma symmetry orbitals are needed. Another problem is that the inorganic systems may proceed by dissociation, so that the process is not concerted, as required for the extended Woodward–Hoffmann rules. The latter complication arises because C—H and C—C bonds are much stronger than metal–carbon bonds, so that dissociation is unlikely in the organic systems. Nevertheless, the current organometallic studies generally attempt to interpret results in terms of symmetry rules and focus on agreement and apparent exceptions to the rules.

A process is symmetry allowed if the migrating group can maintain proper overlap as it moves from the original position to make a new bond with an appropriate empty antibonding orbital, which produces the new species. In addition, one must take into account the concepts of antarafacial and suprafacial movement, which are described in the following examples.

A *1,3 sigmatropic shift* is symmetry allowed but occurs *antarafacially*. This means that the migrating group moves from one side of the molecule to the other. Figure 4.9 shows an orbital picture for such a shift. The orbital diagram on the left in Figure 4.9 shows the unoccupied antibonding π orbital (π^*), which becomes the bonding π orbital in the product. To maintain proper symmetry overlap during the migration, the σ orbital must move from the top to the bottom of the molecule (antarafacially). This movement is viewed as unlikely for organometallic systems and the 1,3 process is not expected to be favorable.

A *1,5 sigmatropic shift* is symmetry allowed and occurs *suprafacially* with the migrating group staying on the same side of the molecule, as shown in Figure 4.10. This process should be easy for organometallic systems.

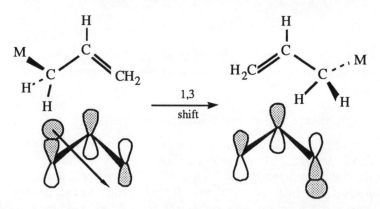

Figure 4.9. A structural and bonding representation of a 1,3 antarafacial sigmatropic shift.

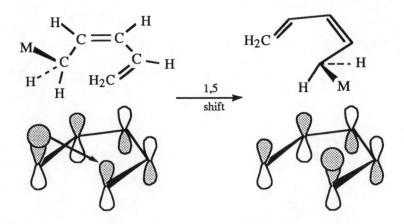

Figure 4.10. A structural and bonding representation of a 1,5 suprafacial sigmatropic shift.

Extensions to larger systems follow in an obvious way and predict that *1,7 shifts should be antarafacial, and 1,9 shifts suprafacial.*

Mingos[54] extended these ideas to π-bonded cyclic systems by using a simple valence bond approach that greatly expanded the applicability of these rules in organometallic systems. Essentially, one draws the valence bond structure and moves electron pairs in the standard way of organic chemistry to obtain the product. If the predicted rearrangement is symmetry allowed and suprafacial, then the fluxional process is expected to have a low energy barrier. For example, with $Fe(CO)_2(\eta^5\text{-}C_5H_5)(\eta^1\text{-}C_5H_5)$ discussed earlier, the pertinent process would be pictured as in Eq. (4.9):

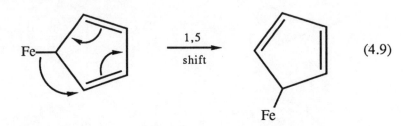

$$(4.9)$$

Note that this corresponds to a 1,5 shift and would be predicted to be of low energy. We called this a 1,2 shift previously, but that is the same as a 1,5 shift for a C_5 ring.

An unexplained exception to the predictions is the $(OC)_5Re(\eta^1\text{-}C_7H_7)$ system,[55] which shows 1,2 (= 1,7) shifts with an energy barrier of ~80 kJ mol^{-1}. The symmetry rules predict that a 1,5 shift should be allowed, as shown in Figure 4.11. A minor 1,5 pathway has been observed with $(\eta^5\text{-}C_5H_5)(CO)_2Re(\eta^1\text{-}C_7H_7)$. Mann has suggested that the reaction may

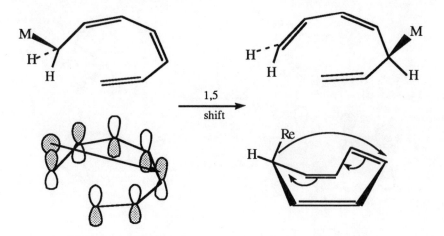

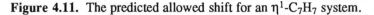

Figure 4.11. The predicted allowed shift for an η^1-C_7H_7 system.

involve homolytic fission of the Re—C bond, or that the rules breakdown because of participation of metal p or d orbitals.

The same approach can be used for higher levels of "hapticity." For example, an η^2-C_6H_6 can undergo a favorable 1,5 shift, but η^4-C_6H_6 requires an unfavorable 1,3 shift, as shown in Figure 4.12.

The η^3-C_7H_7 system is predicted to be easily fluxional by a 1,5 shift, but the η^5-C_7H_7 should be static or highly hindered due to a forbidden 1,3 shift. Both shifts are shown in Figure 4.13. In fact, both of these systems are found to be highly fluxional.

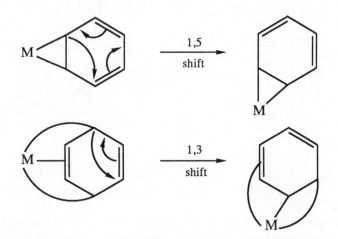

Figure 4.12. The allowed 1,5 shift for η^2-C_6H_6 and the forbidden 1,3 shift for η^4-C_6H_6.

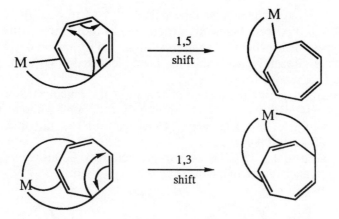

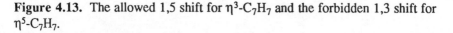

Figure 4.13. The allowed 1,5 shift for η^3-C_7H_7 and the forbidden 1,3 shift for η^5-C_7H_7.

The main exception to this prediction is $[Fe(\eta^5$-$C_7H_7)(CO)_3]^+$ (coalescence temp., $-60°C$) and to a lesser extent $Mn(\eta^5$-$C_7H_7)(CO)_3$ (coalescence temp., $20°C$).[56] The charge dependence of the activation parameters and the rather exceptional deviation from the rules has led to the proposal of a different mechanism involving a charge-separated intermediate instead of a concerted process.

For cyclooctatetraene systems, the prediction is that the η^2-C_8H_8 systems should be static because a 1,7 shift is required. The η^4-C_8H_8 systems should be fluxional by a 1,5 shift, but the η^6-C_8H_8 complexes should be static because a 1,3 shift is required, as shown in Figure 4.14.

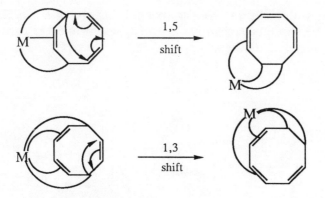

Figure 4.14. The allowed 1,5 shift for η^4-C_8H_8 and the forbidden 1,3 shift for η^6-C_8H_8.

The η^6-C_8H_8 systems that have been studied,[57] $M(\eta^6$-$C_8H_8)(CO)_3$ (M = Cr or W), are fluxional, but with barriers >60 kJ mol^{-1} and with a mixture of shift types. This behavior has been attributed to formation of an η^4-C_8H_8 intermediate. Recently, a 1,5 shift has been identified in $Os(\eta^6$-$C_8H_8)(\eta^4$-1,5-cyclooctadiene).[58]

The general experimental observations are that the neutral complexes of η^2-C_6R_6, η^3-C_7H_7, and η^4-C_8H_8 are highly fluxional, whereas those of η^4-C_6R_6, η^5-C_7H_7, η^2-C_8H_8, and η^6-C_8H_8 are usually static or show high barriers (>80 kJ mol^{-1}).

The 1,5 shift in the η^4-C_8H_8 system predicts specific movements in the remaining ligands on the metal, as shown in Eq. (4.10):

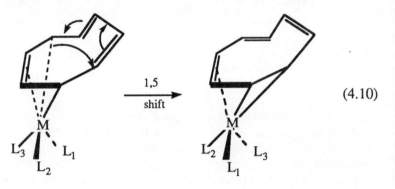

$$\text{(4.10)}$$

These predictions have been tested for $Fe(\eta^4$-$C_8H_8)(CO)_2(i$-PrNC) and are claimed to be confirmed.[59] The system is more complex than ideal because there are two stable structural isomers with *i*-PrNC in positions L_1 or L_2 in the left-hand structure. However, for $Os(\eta^3$-$C_7H_7)(CO)_3(SnPh_3)$, a similar test is not consistent with the predictions and the observations indicate that 1,2 shifts predominate.[60] The latter results seem more consistent with a slip mechanism, possibly involving a *p* orbital on the metal. The slip mechanism has been suggested in other cases and one version is pictured in Scheme 4.5.

Scheme 4.5

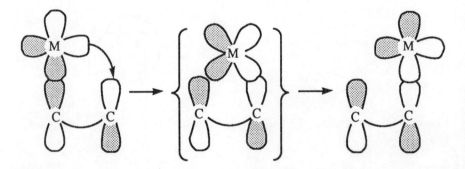

4.4.d   Rotations of π-Bonded Olefins

The bonding in π-bonded olefins consists of donation from the olefin π orbital to a metal σ orbital and back donation from a metal d to the olefin π* orbital, as shown in the following diagram:

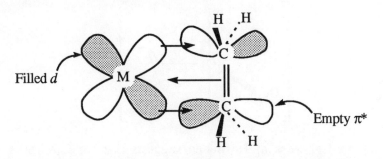

This amounts to at least a partial double bond, and some restricted rotation of the olefin relative to the metal might be expected. The rotation could be about the M–olefin bond or about the C=C axis. These compounds normally have the C=C axis perpendicular to the square plane with L_3M units, and parallel to the trigonal plane with L_4M units. Hoffmann and co-workers[61] have given an extensive theoretical analysis of the bonding and fluxional processes in these systems. They conclude that the structures of the square planar complexes are largely controlled by steric factors, while the trigonal bipyramidal complexes prefer ethylene in the trigonal xy plane because of better overlap with the hybridized d_{xy} orbital compared to the unhybridized d_{xz} orbital.

Cramer et al.[62] observed ethylene rotation in $M(\eta^5\text{-}C_5H_5)(C_2H_4)_2$ (M = Rh, Ir) complexes with the following structure:

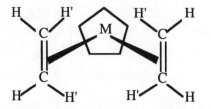

For M = Rh, the H and H' can be observed separately in the nmr spectrum at –20°C, but the peaks collapse to one at 57°C. The analogous C_2F_4 compound shows no collapse of the ^{19}F spectrum up to 100°C. The rotational barrier also is increased by CH_3 groups on the cyclopentadiene ring and by changing from Rh to Ir. These differences can be explained by steric and better back π-bonding factors, but they do not determine whether rotation is about the metal–olefin or C=C bond.

Rotation about the metal–olefin bond has been observed[63] in $Os(PPh_3)_2(CO)(NO)(C_2H_4)$, shown in the following diagram. The two ends of the ethylene are different and the ^{13}C spectrum shows that they are static at $-80°C$, but interconversion is observed as the temperature is raised to $20°C$. Since the $^1H-^{31}P$ coupling is retained, the reaction does not involve dissociation of ethylene.

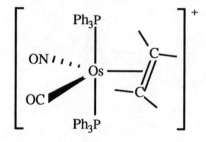

The theoretical work of Hoffmann suggests that the L_4M–ethylene systems should rearrange by a pseudorotation mechanism with the ethylene remaining in the trigonal plane. The $(OC)_4M(ethylene)$ compounds (M = Fe, Ru, Os) show axial–equatorial CO interchange.[64] Caulton and co-workers[65] have suggested that ethylene rotation may be facilitated by interaction with π^* orbitals of axial CO ligands in a transition state such as that in the following diagram. This is used to rationalize the nonfluxional character of systems which lack a CO ligand, such as $(Me_2PhP)_4Ir(C_2H_4)^+$, $(Me_2PhP)_3(CH_3CN)Ir(C_2H_4)^+$, and $(Me_2PhP)_3(CH_3)Ir(C_2H_4)$.

4.4.e   Fluxional Allyl Complexes

The allyl group $(-C_3H_5^-)$ can be bonded to a metal either as $\eta^1-C_3H_5$ or as $\eta^3-C_3H_5$, as shown in the following diagrams. The former is normally observed with nontransition metals and the latter is the predominant form with transition metals.

In the $\eta^1-C_3H_5$ complexes, fluxional behavior involves movement of the metal from one end of the allyl ligand to the other. In $\eta^3-C_3H_5$ systems, the

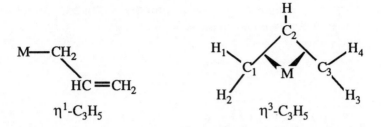

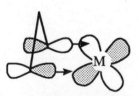

η^1-C_3H_5

η^3-C_3H_5

syn protons (H_1, H_4) and anti protons (H_2, H_3) may interchange, and C_1 and C_3 may be observed to interchange if other ligands make the two sides of the complex different. The bonding[66] and fluxional aspects[67] of these systems have been reviewed.

The η^1-C_3H_5 systems are predicted to be static by the symmetry rules, and this is the case for $(OC)_5Mn(\eta^1$-$C_3H_5)$. However, if the metal is coordinatively unsaturated (<18 valence electrons), then the systems are highly fluxional and are thought to proceed through an η^3-C_3H_5 intermediate, such as in $(Me_2PhP)(NC_5H_4CO_2)Pd(\eta^1$-$C_3H_5)$ and $(Me_2N)_2Ti(\eta^1$-$C_3H_5)$. A similar state can be attained by reversible loss of another ligand, and heterolytic bond cleavage also has been proposed for an Fe complex.[68]

The reason for restricted rotation in the η^3-C_3H_5 complexes can be understood from the following bonding pictures, that depict the "sigma" donation from the ligand on the left and the back donation from the metal to the π^* orbital of the ligand on the right.

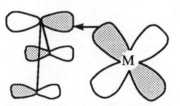

σ donation to metal   π back bonding

Three processes have been suggested for the fluxionality observed in the η^3-C_3H_5 systems:
1. In the olefin rotation mechanism, the intermediate has a bond only with the C_1–C_2 end, and the C_3 is free to rotate and interchange the syn and anti protons.
2. In the allyl flip mechanism, the intermediate is bonded only at C_1 and C_3, and the C_2 does an end-over flip. The syn and anti protons will interchange at both ends of the allyl ligand.
3. In the π–σ–π mechanism, the intermediate is σ bonded at one end (e.g., C_1) in an η^1 fashion, and the ligand is free to rotate about the C_1—C_2 bond. Each event causes syn–anti exchange at one end of the ligand.

The "π–σ–π" mechanism is now commonly accepted. Some of the evidence is from observations on the following system:[69]

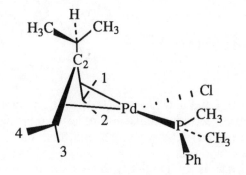

The 3 and 4 protons interchange at the same rate as the methyl protons on the phosphine and as the methyl protons of the 2-isopropyl group. This is expected if the C_2 is moving above and below the P–Pd–Cl plane as in the "π–σ–π" mechanism.

4.4.f   Bridge–Terminal Carbonyl Exchange

The phenomenon of bridge–terminal carbonyl exchange can occur in any dimetal system that can exist in CO-bridged and metal–metal bonded forms, such as Co_2CO_8, shown in Eq. (4.11):

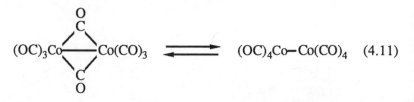

$$(OC)_3Co\overbrace{\underbrace{}}Co(CO)_3 \rightleftharpoons (OC)_4Co-Co(CO)_4 \quad (4.11)$$

Another example[70] is $[(\eta^5\text{-}C_5H_5)Ru(CO)_2]_2$, where exchange is rapid on the nmr time scale down to –100°C.

A more complex and informative process has been observed with $[(\eta^5\text{-}C_5H_5)Fe(CO)_2]_2$, which has bridging and terminal CO ligands and cis and trans isomers, as shown in Figure 4.15.

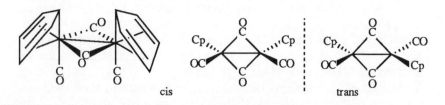

Figure 4.15. The structure and isomers of $[(\eta^5\text{-}C_5H_5)Fe(CO)_2]_2$.

The temperature dependence of the cis–trans equilibrium adds a complication to the nmr analysis. Several detailed studies[71] have shown that

1. Bridge–terminal CO exchange has a lower activation energy in the trans isomer.
2. Bridge–terminal exchange in the cis isomer occurs at the same rate as the cis–trans isomerization.

These observations have been explained by Adams and Cotton[72] by the mechanism as depicted in Scheme 4.6.

Scheme 4.6

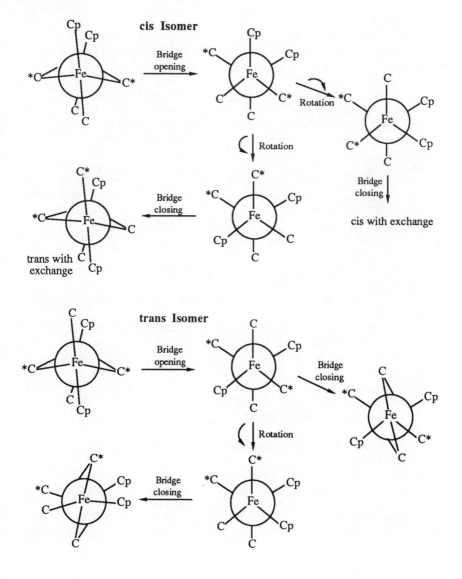

The basis of the mechanism is a nonbridged intermediate, which in the cis case must undergo rotation to cause bridge–terminal exchange and isomerization. But the trans isomer can undergo exchange without rotation. It is necessary to remember that ring closing can only occur with two CO ligands on opposite sides of the molecule. This general mechanism seems to be consistent with observations on other systems. For example, in $(\eta^5\text{-}C_5H_5)_2Fe_2(CO)_3(P(OPh)_3)$ the cis–trans isomerization and bridge–terminal exchange have the same rate because the phosphite ligand cannot occupy a bridging position.[73]

References

1. Wilkins, R. G. *The Study of Kinetics and Mechanism of Reactions of Transition Metal Complexes*; Allyn and Bacon: Boston, 1974; Chapter 7.
2. Basolo, F.; Pearson R. G. *Mechanisms of Inorganic Reactions*, 2nd ed.; Wiley: New York, 1967.
3. Holm, R. H.; O'Connor, M. J. *Prog. Inorg. Chem.* **1971**, *14*, 241.
4. Serpone, N.; Bickley, D. G. *Prog. Inorg. Chem.* **1972**, *17*, 391.
5. *Dynamic Nuclear Magnetic Resonance Spectroscopy*; Jackman, L. M.; Cotton, F. A., Eds.; Academic Press: New York, 1975; Chapters 8–12.
6. Beattie, J. K. *Acc. Chem. Res.* **1971**, *4*, 253.
7. McGarvey, J. J.; Wilson, J. *J. Am. Chem. Soc.* **1975**, *97*, 2531.
8. Ivin, K. J.; Jamison, R.; McGarvey, J. J. *J. Am. Chem. Soc.* **1972**, *94*, 1763; Campbell, L.; McGarvey, J. J.; Samman, N. G. *Inorg. Chem.* **1978**, *17*, 3378.
9. Farina, R. D.; Swinehart, J. H. *Inorg. Chem.* **1972**, *11*, 645.
10. Balahura, R. J.; Lewis, N. A. *Coord. Chem. Rev.* **1976**, *20*, 109.
11. Burmeister, J. L. *Coord. Chem. Rev.* **1968**, *3*, 225.
12. Brasch, N. E.; Buckingham, D. A.; Clark, C. R.; Finnie, K. S. *Inorg. Chem.* **1989**, *28*, 4567.
13. Basolo, F.; Baddley, W. H.; Weidenbaum, K. J. *J. Am. Chem. Soc.* **1966**, *88*, 1576.
14. Jackson, W. G.; Lawrance, G. A.; Lay, P. A.; Sargeson, A. M. *Aust. J. Chem.* **1982**, *35*, 1562.
15. Damrauer, L.; Milburn, R. M. *J. Am. Chem. Soc.* **1971**, *93*, 6481.
16. Dalzell, B. C.; Eriks, K. *J. Am. Chem. Soc.* **1971**, *93*, 4298.
17. Gamsjäger, H.; Milburn, R. K. *Adv. Inorg. Bioinorg. Mech.* **1983**, *2*, 317.
18. Palmer, D. A.; Kelm, H. *J. Inorg. Nucl. Chem.* **1978**, *40*, 1095.
19. Odell, A. L.; Olliff, R. W.; Rands, D. B. *J. Chem. Soc., Dalton Trans.* **1972**, 752.
20. Lethbridge, J. W.; Glasser, L. S. D.; Taylor, H. F. W. *J. Chem. Soc. A* **1970**, 1862.
21. Lawrance, G. A.; Stranks, D. R. *Inorg. Chem.* **1977**, *16*, 929.
22. Evilia, R. F.; Young, D. C.; Reilley, C. N. *Inorg. Chem.* **1971**, *10*, 433; Ho, F. F.-L.; Reilley, C. N. *Anal. Chem.* **1969**, *41*, 1835.
23. Wilkins, R. G.; Williams, M. J. G. *J. Chem. Soc.* **1957**, 1763.
24. Basolo, F.; Hayes, J. C.; Neumann, H. M. *J. Am. Chem. Soc.* **1954**, *76*, 3807.

25. Lawrance, G. A.; Stranks, D. R. *Inorg. Chem.* **1978**, *17*, 1804; Ibid. **1977**, *16*, 929.
26. Beattie, J. K.; Binstead, R. A.; West, R. J. *J. Am. Chem. Soc.* **1978**, *100*, 3046.
27. McGarvey, J. J.; Lawthers, I.; Heremans, K.; Toftlund, H. *Inorg. Chem.* **1990**, *29*, 252.
28. Vanquickenborne, L. G.; Pierloot, K. *Inorg. Chem.* **1981**, *20*, 3673.
29. Vanquickenborne, L. G.; Pierloot, K. *Inorg. Chem.* **1984**, *23*, 1471.
30. Wilkins, R. G. *The Study of Kinetics and Mechanism of Reactions of Transition Metal Complexes*; Allyn and Bacon: Boston, 1974; p. 350.
31. Holm, R. H. In *Dynamic Nuclear Magnetic Resonance Spectroscopy*; Jackman, L. M.; Cotton, F. A., Eds.; Academic Press: New York, 1975; p. 317.
32. Musher, J. I. *J. Chem. Educ.* **1974**, *51*, 94.
33. Eaton, S. S.; Hutchinson, J. R.; Holm, R. H. *J. Am. Chem. Soc.* **1972**, *94*, 6411.
34. Eaton, S. S.; Eaton, G. R.; Holm, R. H. *J. Am. Chem. Soc.* **1973**, *95*, 1116.
35. Pignolet, L. H.; Duffy, D. J.; Que, L., Jr. *J. Am. Chem. Soc.* **1973**, *95*, 295; Palazzotto, M. C.; Duffy, D. J.; Edgar, B. L.; Que, L., Jr.; Pignolet, L. H. *J. Am. Chem. Soc.* **1973**, *95*, 4537.
36. Rodger, A.; Johnson, B. F. G. *Inorg. Chem.* **1988**, *27*, 3061.
37. Keppert, D. *Prog. Inorg. Chem.* **1977**, *23*, 1.
38. Meakin, P.; Jesson, J. P. *J. Am. Chem. Soc.* **1973**, *95*, 7272.
39. Berry, R. S. *J. Chem. Phys.* **1960**, *32*, 933.
40. Ugi, I.; Marquarding, D.; Klusacek, H.; Gillespie, P. *Acc. Chem. Res.* **1971**, *4*, 288.
41. Strich, A. *Inorg. Chem.* **1978**, *17*, 942.
42. Meakin, P.; Muetterties, E. L.; Jesson, J. P. *J. Am. Chem. Soc.* **1972**, *94*, 5271.
43. Mann, B. E. In *Comprehensive Organometallic Chemistry*; Abel, E. W.; Stone, F. G. A.; Wilkinson, G., Eds.; Pergamon Press: London, 1982; Vol. 3, p. 89.
44. Mann, B. E. *Chem. Soc. Rev.* **1986**, *15*, 167.
45. Meakin, P.; Muetterties, E. L.; Jesson, J. P. *J. Am. Chem. Soc.* **1972**, *94*, 5271.
46. Sheline, R. K.; Mahnke, H. *Angew. Chem., Int. Ed.* **1975**, *14*, 314.
47. Mahnke, H.; Clark, R. J.; Rosanske, R.; Sheline, R. K. *J. Chem. Phys.* **1973**, *60*, 2997.
48. Speiss, H. W.; Grosescu, R.; Haeberlen, U. *Chem. Phys.* **1974**, *6*, 226.
49. Hanson, B. E.; Whitmire, K. H. *J. Am. Chem. Soc.* **1990**, *112*, 974.
50. Burdett, J. K.; Gryzbowski, J. M.; Poliakoff, M.; Turner, J. J. *J. Am. Chem. Soc.* **1976**, *98*, 5728.
51. Bennett, M. J.; Cotton, F. A.; Davison, A.; Faller, J. W.; Lippard, S. J.; Morehouse, S. M. *J. Am. Chem. Soc.* **1966**, *88*, 4371; Piper, T. S.; Wilkinson, G. *J. Inorg. Nucl. Chem.* **1956**, *3*, 104.
52. Cotton, F. A.; Musco, A.; Yagupsky, G. *J. Am. Chem. Soc.* **1967**, *89*, 6136.
53. Woodward, R. B.; Hoffmann, R. *J. Am. Chem. Soc.* **1965**, *87*, 2511.

54. Mingos, D. M. P. *J. Chem. Soc., Dalton Trans.* **1977**, 602.
55. Heinekey, D. M.; Graham, W. A. G. *J. Am. Chem. Soc.* **1982**, *104*, 915; Ibid. **1979**, *101*, 6115; *J. Organomet. Chem.* **1982**, *232*, 335.
56. Whitesides, T. H.; Budnik, R. A. *Inorg. Chem.* **1976**, *15*, 874.
57. Gibson, J. A.; Mann, B. E. *J. Chem. Soc., Dalton Trans.* **1979**, 1021.
58. Grassi, M.; Mann, B. E.; Spencer, C. M. *J. Chem. Soc., Chem. Commun.* **1985**, 1169.
59. Hails, M. J.; Mann, B. E.; Spencer, C. M. *J. Chem. Soc., Dalton Trans.* **1985**, 693.
60. Takats, J.; Kiel, G.-Y. *Organometallics* **1987**, *6*, 2009.
61. Albright, T. A.; Hoffmann, R.; Thibeault, J. C.; Thorn, D. L. *J. Am. Chem. Soc.* **1979**, *101*, 3801.
62. Cramer, R.; Kline, J. B.; Roberts, J. D. *J. Am. Chem. Soc.* **1969**, *91*, 2519.
63. Segal, J. A.; Johnson, B. F. G. *J. Chem. Soc., Dalton Trans.* **1975**, 677.
64. Takats, J.; Burke, M. R. *J. Am. Chem. Soc.* **1983**, *105*, 4092.
65. Lundquist, E. G.; Folting, K.; Streig, W. E.; Huffman, J. C.; Eisenstein, O.; Caulton, K. G. *J. Am. Chem Soc.* **1990**, *112*, 855.
66. Mingos, D. M. P. In *Comprehensive Organometallic Chemistry*; Abel, E. W.; Stone, F. G. A.; Wilkinson, G., Eds.; Pergamon Press: London, 1982; Vol. 3, pp. 60–61.
67 Vrieze, K.; van Leeuwen, P. W. N. M. *Prog. Inorg. Chem.* **1971**, *14*, 1; Clarke, H. L. *J. Organomet. Chem.* **1974**, *80*, 155; Tsutsui, M.; Courtney, A. *Adv. Organomet. Chem.* **1977**, *16*, 241; Anderson, G. K.; Cross, R. J. *Chem. Soc. Rev.* **1980**, *9*, 185.
68. Rosenblum, M.; Waterman, P. *J. Organomet. Chem.* **1981**, *206*, 197.
69. van Leeuwen, P. W. N. M.; Praat, N. A. P.; van Diepen, M. *J. Organomet. Chem.* **1971**, *29*, 433.
70. Bullit, J. G.; Cotton, F. A.; Marks, T. J. *J. Am. Chem. Soc.* **1970**, *92*, 2155.
71. Adams, R. D.; Cotton, F. A. *J. Am. Chem. Soc.* **1973**, *95*, 6589; Gansow, O. A.; Burke, A. R.; Vernon, W. D. *J. Am. Chem. Soc.* **1972**, *94*, 2550.
72. Adams, R. D.; Cotton, F. A. In *Dynamic Nuclear Magnetic Resonance Spectroscopy*, Jackman, L. M.; Cotton, F. A., Eds.; Academic Press: New York, 1975; Chapter 12.
73. Cotton, F. A.; Kruczynski, L.; White, A. J. *Inorg. Chem.* **1974**, *13*, 1402.

5

Reaction Mechanisms of Organometallic Systems

5.1 LIGAND SUBSTITUTION REACTIONS

The general principles discussed in Chapter 3 also apply to reactions of organometallic complexes. Because these systems do not have a wide range of structurally similar complexes with different metal atoms for comparative studies across the Periodic Table, comparisons are usually made down a particular group. However, there is a wide range of ligands available for studies of entering and leaving group effects. This area has been the subject of several recent reviews.[1-3] A major difference from the systems discussed in Chapter 3 is that many of these complexes are soluble in organic solvents, including hydrocarbons. This can minimize the complicating factor of solvent coordination, but these solvents often have quite low dielectric constants so that various types of preassociation are more probable.

5.1.a Metal Carbonyls

The metal carbonyl family of compounds is typical of the range of structures and reactivities of organometallic complexes. The rate of CO exchange was examined in early studies, and this work is the subject of a recent review.[4] The order of reaction rates is as follows:

$$V(CO)_6 > Ni(CO)_4 > Mo(CO)_6 > Cr(CO)_6 >$$
$$W(CO)_6 > Fe(CO)_5 > Mn_2(CO)_{10}$$

Where the rate laws have been determined, the reaction is first-order in the metal carbonyl and zero-order in [CO]. This implies a **D** mechanism since noncoordinating solvents eliminate the possibility of a coordinated solvent intermediate.

The main exception to the preceding generalization is $V(CO)_6$, which has an I_a mechanism for PR_3 substitution reactions.[5] This compound is unique in that it is the only 17-electron metal carbonyl and also by far the most substitution labile of this family. Some kinetic results for substitution on $V(CO)_6$ in hexane are given in Table 5.1.

Table 5.1. Rate Constants (25°C) and Activation Parameters for PR_3 Substitution on $V(CO)_6$ in Hexane

Entering Group	Cone Angle	k $(M^{-1} s^{-1})$	ΔH^* $(kcal\,mol^{-1})$	ΔS^* $(cal\,mol^{-1}\,K^{-1})$
PMe_3	118	132	7.6	−23.4
$P(n\text{-Bu})_3$	132	50.2	7.6	−25.2
$P(OMe)_3$	107	0.70	10.9	−22.6
$P(Ph)_3$	145	0.25	10.0	−27.8

The substitution rates have rather low ΔH^*, and the negative ΔS^* values are typical of an associative process. The rates for various entering groups correlate with the basicity rather than the size as measured by the cone angle. It has been suggested that formation of a 19-electron associative intermediate is much more favorable than a 20-electron intermediate from an 18-electron starting material.

Quantitative studies of CO exchange have been rather limited because of experimental difficulties, especially the low solubility of CO in most solvents. This is illustrated by work on $Ni(CO)_4$. The original study was incorrect because much of the exchange was occurring in the gas phase, where it is faster than in solution. Subsequent work[6] has shown that the CO exchange and PPh_3 substitution occur at the same rate as expected for a dissociative mechanism (k = 2.1 x 10^{-2} M^{-1} s^{-1} (30°C), $\Delta H^* = 24$ kcal mol^{-1}, $\Delta S^* = 13.1$ cal mol^{-1} K^{-1}). It seems surprising that a species with an initially small coordination number such as $Ni(CO)_4$ would react by a **D** mechanism, but it appears that 20-electron associative intermediates or transition states are normally less stable than the 16-electron dissociative ones because the extra electron pair must go into an antibonding orbital.

The exchange of CO on $Fe(CO)_5$ was originally reported to show two rates attributed to the axial and equatorial CO ligands. Of course, this is inconsistent with the very rapid fluxionality of $Fe(CO)_5$ discussed in Chapter 4. Later studies report[7] only one observable rate of CO exchange. The CO exchange on $Fe(CO)_5$ has proved to be too slow to measure relative to thermal decomposition. However, Basolo and co-workers[8] have estimated the rate for $Fe(CO)_5$ based on measured rates for PR_3 substitution in $Os(CO)_5$ and $Ru(CO)_5$.[9] The results are given in Table 5.2. The reactivity pattern down the group is similar to that for Group VI, discussed later, but $Fe(CO)_5$ is still surprisingly unreactive. Basolo has suggested that this could be because the $\{Fe(CO)_4\}$ dissociative intermediate is high-spin, consistent with observations of Poliakoff and Turner,[10] and therefore the process is spin-forbidden. But a spin-allowed dissociation to the higher energy spin-paired species is possible.

Table 5.2. Rate Constants (50°C) and Activation Parameters for PR_3
Substitution on Group V Metal Carbonyls in Decalin

Compound	k (s^{-1})	ΔH^* (kcal mol^{-1})	ΔS^* (cal mol^{-1} K^{-1})
$Fe(CO)_5$	(6×10^{-11})	40	18
$Ru(CO)_5$	3.0×10^{-3}	27.6	15.2
$Os(CO)_5$	4.9×10^{-8}	30.6	1.33

The Group VI $M(CO)_6$ compounds have been most widely studied with regard to CO exchange and substitution. The substitution reactions usually show a two-term pseudo-first-order rate constant, as given in Eq. (5.1):

$$k_{exp} = k_1 + k_2 [L] \qquad (5.1)$$

The k_1 term is assigned to a **D** mechanism and the k_2 pathway to an I_a mechanism. Some results for CO exchange and $P(n\text{-}C_4H_9)_3$ substitution are given in Table 5.3. Clearly, the kinetic parameters for exchange and k_1 are quite similar, as expected if CO dissociation is rate controlling for both reactions. The parallel between f_{M-C} and ΔH^* and ΔH_1^* is expected for a dissociative process.

Any explanation for the lack of a smooth trend in ΔH^* down the group first requires a decision as to which member is out of line. It has been argued by King[11] that $W(CO)_6$ has an especially strong M—C bond because the lanthanide contraction makes the covalent radius of W smaller than expected (Cr, 1.25 Å; Mo, 1.36 Å; W, 1.37 Å), with the result that the $5d$ and maybe $4f$ orbitals give much better back π bonding. A recent theoretical study[12] indicates that this argument is overly simplistic. Back π bonding is

Table 5.3. Activation Parameters for CO Exchange and $P(n\text{-}C_4H_9)_3$
Substitution on $M(CO)_6$ in Decalin^a

	Exchange		Substitution				
Compound	ΔH^*	ΔS^*	ΔH_1^*	ΔS_1^*	ΔH_2^*	ΔS_2^*	$f_{M-C}{}^b$
$Cr(CO)_6$	38.7	18.5	40.2	22.6	25.5	−14.6	2.08
$Mo(CO)_6$	30.2	−0.4	31.7	6.7	21.7	−14.9	1.96
$W(CO)_6$	39.8	11.0	39.9	13.8	29.2	−6.9	2.36

^{a} Activation enthalpies and entropies in kcal mol^{-1} and cal mol^{-1} K^{-1},
   respectively; original sources are given in reference 12.
^{b} Force constant for the M—C bond in mdyn Å^{-1}.

best for the first row elements, and repulsions between d electrons and the CO ligands increase down the series but are compensated by relativistic stabilization effects, which are especially important for the third row. The theory indicates that the balance of these effects can account for the reactivity pattern in both Group VI and Group V.

For phosphine substitution,[13] $\Delta V_1^* = 15$ cm^3 mol^{-1} for $Cr(CO)_6$ in cyclohexane and 10 cm^3 mol^{-1} for $Mo(CO)_6$ in isooctane. The values are positive, as expected for a dissociative process. The ΔV_2^* of -10 cm^3 mol^{-1} for $W(CO)_6$ is negative, as expected for an associative process. The negative values of ΔS_2^* in Table 5.3 are consistent with an associative mechanism. The parallel between ΔH_1^* and ΔH_2^* indicates a significant amount of bond breaking in the associative transition state.

5.1.b   Substituted Metal Carbonyls, $M_m(CO)_nX_x$

The CO exchange and substitution reactions on these systems are usually discussed in terms of the cis and trans effects of the heteroligand X and steric factors. The rate laws are similar to that described for the $M(CO)_6$ systems [Eq. (5.1)], but the k_1 or k_2 path may dominate, depending on the system.

The $Mn(CO)_5X$ (X = Cl or Br) systems have been the subject of a number of studies. Wojcicki and Basolo[14] originally reported that cis-CO exchange was much faster than trans-CO exchange. Later work[15] indicated that the two rates were within a factor of less than two of each other. However, the recent study of Atwood and Brown[16] finds that cis-CO exchange is more than 10 times faster than trans-CO exchange in $Mn(CO)_5Br$ and $Re(CO)_5Br$. Exchange into the trans position is proposed to occur subsequent to cis exchange by a fluxional process in the intermediate. These results have been discussed in Chapter 1 with regard to the principle of microscopic reversibility.

A recurrent problem in this area and for other substitution reactions is the separation of the electronic and steric effects of the nonreacting ligands. In organometallic systems, the steric effect is usually measured by the Tolman cone angle (θ), but the electronic effect is a mixture of σ-donor and π-acceptor abilities of the nonreacting ligands. Chen and Poë[17] have studied the reaction

$$Ru(CO)_4L + L' \xrightarrow{\ k_1\ } Ru(CO)_3(L)(L') + CO \qquad (5.2)$$

and used the ^{13}C chemical shift (δ) in $Ni(CO)_3L$ as a measure of the electronic factor for L to correlate the observations according to Eq. (5.3):

$$\log k_1 = \alpha + \beta_L\, \delta + \gamma_L\, \theta \qquad (5.3)$$

They also found that this approach correlates substitution kinetics for several other substituted metal carbonyls and for a methyl migration reaction.

The heteroligand X can have special properties that influence the substitution lability and mechanism. For example, the $(\eta^5\text{-}C_5H_5)M(CO)_2$ compounds (M = Co, Rh, or Ir) react by an I_a mechanism.[18] This has been explained[19] as "slippage" of the $(\eta^5\text{-}C_5H_5)$ to an $(\eta^3\text{-}C_5H_5)$ form in the transition state, thereby liberating a metal orbital to accept an electron pair from the entering group. A similar argument has been used by Basolo and co-workers[20] to explain the unusually high associative substitution lability of $(\eta^5\text{-indenyl})Rh(CO)_2$, which is about 10^8 times more reactive than the $(\eta^5\text{-}C_5H_5)$ analogue. It is proposed that the intermediate formed by slippage is stabilized by the gain in resonance energy in the six-membered ring, as shown in Scheme 5.1.

Scheme 5.1

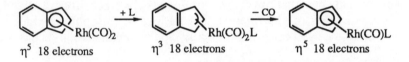

η^5 18 electrons η^3 18 electrons η^5 18 electrons

The danger in making generalizations in these systems is shown by recent CO exchange studies[21] on $(\eta^5\text{-}C_5R_n)_2V(CO)$ compounds. As expected, $(\eta^5\text{-}C_5H_5)_2V(CO)$ and $(\eta^5\text{-}C_5(CH_3)_5)_2V(CO)$ are quite labile and react by an I_a mechanism (k_2). This is consistent with either the 17- to 19- electron intermediate formation or slippage of the ring. The unexpected fact is that derivatives of $(\eta^5\text{-}C_5H_7)_2V(CO)$ are much more inert and show a significant dissociative reaction pathway (k_1). Some data are given in Table 5.4. Strangely, the mixed complex $(\eta^5\text{-}C_5H_5)(\eta^5\text{-}C_5H_7)V(CO)$ is just slightly more reactive than $(\eta^5\text{-}C_5H_7)_2V(CO)$ and quite different from $(\eta^5\text{-}C_5H_5)_2V(CO)$. Basolo has attributed this great difference in reactivity largely to electronic factors related to the poorer donation of electrons to the metal from $\eta^5\text{-}C_5H_7$ than from $\eta^5\text{-}C_5H_5$. There is some evidence for this in

Table 5.4. Rate Constants (60°C) and Activation Parameters for CO Exchange on $(\eta^5\text{-}C_5R_n)_2V(CO)$ Systems in Decalin[a]

Compound	k_1	ΔH_1^*	ΔS_1^*	k_2	ΔH_2^*	ΔS_2^*
$(\eta^5\text{-}C_5Me_5)_2V(CO)$	~10^{-4}			2.6×10^2	8.9	−21
$(\eta^5\text{-}C_5H_5)(\eta^5\text{-}C_5H_7)V(CO)$	3×10^{-4}			5.7×10^{-3}		
$(\eta^5\text{-}C_5H_7)_2V(CO)$	8×10^{-6}	28.1	2	3.8×10^{-3}	22.7	−2

[a] k_1 (s^{-1}), k_2 ($M^{-1}\,s^{-1}$), ΔH^* (kcal mol^{-1}), and ΔS^* (cal mol^{-1} K^{-1}).

the CO stretching frequencies (v_{CO}), which are 1959 cm^{-1} in the former and 1881 cm^{-1} in the latter. The lower value indicates more back donation into the π^* orbital of CO and a stronger M—C bond. In any case, it is fair to say that these reactivity differences were quite unexpected.

Nitrosyl ligands tend to favor an associative mechanism. For example, $Co(NO)(CO)_3$ and $Fe(NO)_2(CO)_2$ are isoelectronic with $Ni(CO)_4$ but have associative substitution mechanisms.[22] Basolo has suggested that if the NO is formally regarded as NO$^+$, then its stronger π-acceptor ability relative to CO can be rationalized, and NO$^+$ will have a greater tendency to remove electron density from the metal and favor an associative mechanism.

5.1.b.i   Cis Labilizing Effect

The cis labilizing effect has been discussed previously, and the order of cis labilizing influence is opposite to that of the trans effect, with π-donor ligands being the most cis labilizing. Recent work[23] of Darensbourg and co-workers indicates that oxygen-donor ligands are especially effective for the cis labilization of CO ligands, and the use of phosphine oxides and acetate ion in the synthesis of specifically labeled metal carbonyls has been especially useful.

Rossi and Hoffmann[24] suggested that poor or non π acceptors would prefer the equatorial site in a square pyramidal dissociative intermediate for d^6 metal complexes. This prediction has been exploited in stereoselective labeling work, since it predicts that such heteroligands will minimize the amount of scrambling due to fluxionality in the intermediate. On the other hand, if the heteroligand is a better σ donor and π acceptor than CO, then the heteroligand should favor the axial position in the square pyramid and should also minimize fluxionality. This approach has been used[25] with CS as the heteroligand for the synthesis in Eq. (5.4):

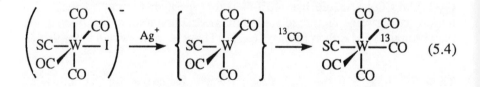

$$\tag{5.4}$$

5.1.b.ii   Heteroligand Replacement

For $M(CO)_5L$ systems, the kinetic order for replacement of the leaving group L in a first-order process is generally as follows:

$$py > AsPh_3 > CO \approx PPh_3 > P(OPh)_3 > P(OMe)_3 > P(n\text{-}Bu)_3$$

High reactivity may be associated with poor π-acceptor ability, as for py and AsPh$_3$, or with poor σ-donor ability, as for CO. The phosphites are considered to be better π acceptors than the phosphines, but steric factors enter as a compensating factor. This type of dichotomy pervades the interpretation of reactivity patterns for these systems.

For *cis*-Mo(CO)$_4$L$_2$ complexes, steric effects seem to dominate the substitution reactions. For L = PPh$_3$, the rate of L displacement by CO is ~200 times larger than that for L = PMePh$_2$. The ground-state structures[26] show that the P—Mo—P angle is distorted from 90° to 104.6° in the former and only to 92.5° in the latter. The distortion of the structure is relieved in the transition state for a dissociative process. It should be noted that the Mo—P bond lengths are very similar, 2.577 Å for PPh$_3$ and 2.555 Å for PMePh$_2$.

The replacement of amine in M(CO)$_5$(NHR$_2$) by phosphines is catalyzed by phosphine oxides. This has been ascribed[27] to a preassociation phenomenon involving hydrogen bonding of the phosphine oxide to the amine hydrogen, thereby weakening the M—N bond. With OP(n-Bu)$_3$ and for M = Mo and R = NHC$_5$H$_{10}$, the kinetics show a saturation effect with the concentration of oxide. The equilibrium constant (hexane at 34.5°C) for formation of the adduct with the following proposed structure is 600 M^{-1}.

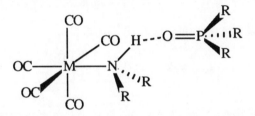

This is a particular case of general Lewis base catalysis of a substitution reaction. It may be troublesome for other studies because of the ease with which some phosphines are oxidized to phosphine oxides. This type of preassociation and catalysis will be favored by the low-polarity solvents used in this area and would probably be unobserved in more polar and hydroxylic solvents.

Chelate ring-opening processes have also been studied in these systems. Graham and Angelici[28] studied the reaction of W(CO)$_4$(bpy) with phosphites and found that the pseudo-first-order rate constant is given by Eq. (5.5):

$$k_{exp} = k_1 + k_2 \text{ [phosphite]} \tag{5.5}$$

Memerring and Dobson[29] noted that the k_1 values given by Graham and Angelici were not the same for different phosphites, and this is not consistent with a **D** mechanism for the k_1 path. They suggested that the reaction was proceeding through a ring-opened intermediate, as shown in Scheme 5.2, and that the apparent k_1 values were really a composite of these two processes that resulted from an incomplete rate law in the original work. The A mechanism of the k_2 path is not shown in Scheme 5.2.

Scheme 5.2

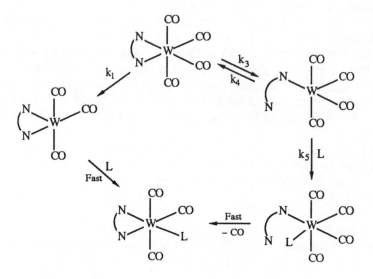

The experimental rate constant is given by Eq. (5.6):

$$k_{exp} = k_1 + k_2\,[L] + \frac{k_3\,k_5\,[L]}{k_4 + k_5\,[L]} \qquad (5.6)$$

A detailed analysis confirmed this rate law and gave consistent k_1 values. More recent work is given by Dobson and co-workers[30] for other chelates and for systems in which the chelate is actually displaced. These reactions generally conform to the preceding mechanism. Recently, the volumes of activation have been measured[31] for displacement of sulfur-bonding chelates from $Cr(CO)_4(S\frown S)$ and the values of 14 cm^3 mol^{-1} are consistent with the **D** ring-opening process.

A coordinatively unsaturated species,[32] $W(CO)_3(PCy_3)_2$, has been found to undergo very rapid substitution,[33] with rate constants in the range of 10^3 to 10^6 M^{-1} s^{-1}, depending on the steric bulk of the entering group. The reactant is stabilized by an "agostic bond" to an H from a Cy ring. If these H atoms are replaced by D atoms, the rate of substitution of $P(OMe)_3$ increases by a factor of 1.15 at 25°C in toluene.

5.1.c   Metal–Metal Bonded Carbonyls

Metal–metal bonded carbonyl systems have received considerable attention and have been the subject of some controversy in recent years. The reactions may proceed by a normal substitution reaction or by homolytic cleavage of the M—M bond, as shown in Scheme 5.3, to give radicals that then undergo rapid substitution before recombining to form the products. The homolysis is a possibility because of the weakness of the M—M bonds.

Scheme 5.3

$$(OC)_nM{-}M(CO)_n + L \longrightarrow (OC)_nM{-}M(CO)_{n-1}L + CO$$

$$(OC)_nM{-}M(CO)_n \underset{}{\overset{\text{Homolysis}}{\rightleftharpoons}} 2\{\cdot M(CO)_n\}$$

$$\{\cdot M(CO)_n\} \longrightarrow \{\cdot M(CO)_{n-1}L\} + CO$$

$$\{\cdot M(CO)_n\} + \{\cdot M(CO)_{n-1}L\} \longrightarrow (OC)_nM{-}M(CO)_{n-1}L$$

The first study on $Mn_2(CO)_{10}$ by Wawersik and Basolo[34] indicated a dissociative mechanism for PR_3 substitution because the rate was inhibited by free CO. However, later work by Poë and co-workers[35] claimed to find no CO inhibition and proposed the homolytic cleavage mechanism in Scheme 5.3. It is important to note that these were primarily studies of the decomposition of $Mn_2(CO)_{10}$ either in the absence or presence of O_2, mainly in decalin at temperatures of 115°C to 180°C. Poë and co-workers found that the rate is half-order in $[Mn_2(CO)_{10}]$ under an argon atmosphere and proposed the mechanism in Eq. (5.7):

$$Mn_2(CO)_{10} \underset{k_{-1}}{\overset{k_1}{\rightleftharpoons}} 2\{Mn(CO)_5\} \overset{k_2}{\longrightarrow} \text{Products} \qquad (5.7)$$

If the dimetal and monometal species are represented by M_2 and M, respectively, and a steady state is assumed for M, then

$$2 k_1 [M_2] - 2 k_{-1} [M]^2 - k_2 [M] = 0 \qquad (5.8)$$

and

$$\frac{d[M_2]}{dt} = k_1 [M_2] - k_{-1} [M]^2 = \frac{k_2 [M]}{2} \qquad (5.9)$$

The first equation is quadratic and can be solved for [M] in terms of $[M_2]$ to give

$$[M] = \frac{k_2}{4 k_{-1}} \left[-1 \pm \left(1 + \frac{16 k_{-1} k_1}{k_2^2} [M_2] \right)^{1/2} \right]$$

$$= \frac{k_2}{4 k_{-1}} \left[-1 + (1 + a)^{1/2} \right] \qquad (5.10)$$

where the positive root is chosen because [M] must be positive. If Eq. (5.10) is multiplied and divided by $\left[1 + (1 + a)^{1/2} \right]$, then one obtains

$$[M] = \frac{k_2}{4\,k_{-1}}\left(\frac{a}{1+(1+a)^{1/2}}\right) \tag{5.11}$$

Substitution for [M] in Eq. (5.9) gives

$$-\frac{d\,[M_2]}{dt} = \frac{(k_2)^2}{8\,k_{-1}}\left(\frac{a}{1+(1+a)^{1/2}}\right) \tag{5.12}$$

If $a^{1/2} \gg 1$, which requires that $a \gg 1$, then Eq. (5.12) yields

$$-\frac{d\,[M_2]}{dt} = \frac{(k_2)^2}{8\,k_{-1}}(a)^{1/2} = \frac{k_2}{2}\left(\frac{k_1\,[M_2]}{k_{-1}}\right)^{1/2} \tag{5.13}$$

Therefore, the rate predicted by this mechanism can be half-order in $[M_2]$. It remains questionable as to whether the assumption concerning the magnitude of a is reasonable. More recent work[36] has shown that the recombination step approaches a diffusion controlled rate ($k_{-1} \approx 10^9\ M^{-1}\ s^{-1}$).

Recent studies have questioned the observations of Poë and co-workers and especially their relevance to the mechanism for CO substitution on $Mn_2(CO)_{10}$. Sonnenberger and Atwood[37] studied $(OC)_5Mn$—$Re(CO)_5$ + PR_3 and found no $Mn_2(CO)_9(PR_3)$ or $Re_2(CO)_9(PR_3)$ products. They expected the latter to form by radical recombination, but this expectation has been questioned by Poë on the basis of probable lifetimes and reactivities of the $\{\bullet M(CO)_5\}$ intermediate. Atwood has argued that the time dependence of the product distribution also is inconsistent with a radical mechanism, since $(OC)_5Mn$—$Re(CO)_4(PR_3)$ and $(R_3P)(OC)_4Mn$—$Re(CO)_5$ increase in concentration and then decay as the disubstituted species is formed. Meutterties and co-workers[38] found that $^{185}Re_2(CO)_{10}$ and $^{187}Re_2(CO)_{10}$ show no isotopic scrambling under thermal decomposition conditions, even when CO is added to suppress the decomposition. They also found that $Mn_2(^{12}CO)_{10}$ and $Mn_2(^{13}CO)_{10}$ show no CO scrambling in octane at 120°C over time periods in which there is substantial exchange with free CO. In addition, the half-times for CO exchange and PPh₃ substitution are 45 and 46 minutes, respectively, indicative of the same rate-controlling process for both reactions. The various details of this problem are discussed in an exchange of notes between Poë[39] and Atwood.[40]

5.1.d   Radical Pathways for Ligand Substitution

Although the question of radical pathways in the metal–metal bonded systems remains doubtful, there are examples of organometallic radical reactions, and they may be more prevalent than originally expected.

Absi-Halabi and Brown[41] found that substitution on Cl_3Sn—$Co(CO)_4$ shows properties characteristic of a radical process. The reaction is catalyzed by light and inhibited by radical traps such as O_2 and galvinoxyl. The proposed process is given in Scheme 5.4, followed by an assortment of radical recombination reactions and electron transfers to the reactant.

Scheme 5.4

$$Cl_3Sn—Co(CO)_4 + L \longrightarrow Cl_3(L)Sn—Co(CO)_4$$

$$Cl_3(L)Sn—Co(CO)_4 \longrightarrow \{Cl_3SnL\bullet\} + \{\bullet Co(CO)_4\}$$

$$\{\bullet Co(CO)_4\} + L \longrightarrow \{\bullet Co(CO)_3(L)\} + CO$$

Byers and Brown[42] observed that $HRe(CO)_5$ is inert toward phosphine substitution at 60°C in hexane in the dark under an N_2 atmosphere. However, any sort of radical initiator gave complete substitution in a few hours, and the reactive species was proposed to be $\{\bullet Re(CO)_5\}$. On the other hand, $HMn(CO)_5$ appears to undergo normal substitution[43] without evidence for a radical process. However, Sweany and Halpern[44] reported that the hydrogenation of $(Ph)(Me)C=CH_2$ by $HMn(CO)_5$ proceeds by a radical pathway, based on the observation of chemically induced dynamic nuclear polarization (CIDNP) in the proton nmr of the methyl styrene. The radicals initially formed by hydrogen atom abstraction by methyl styrene from $HMn(CO)_5$ can either recombine or diffuse apart and undergo further reaction. Bullock and Samsel[45] have suggested a similar mechanism for the reactions of several metal carbonyl hydrides with α-cyclopropylstyrene. This work also provides relative rates for H-atom abstraction and cyclopropyl ring opening by the organic radical.

The mechanism of substitution on the 17-electron radical species has been the subject of recent work. For $\{\bullet Mn(CO)_5\}$, Herrinton and Brown[46] have shown that the substitution is a bimolecular process with the rate constants for different entering groups given in Table 5.5. It also was concluded from this study that decomposition of $\{\bullet Mn(CO)_5\}$ by loss of CO must have a rate constant <90 s^{-1}.

Table 5.5. Rate Constants (24°C) for the Reaction of $\{\bullet Mn(CO)_5\}$ with Various Entering Groups in Hexane

Entering Group	k (M^{-1} s^{-1})
PPh$_3$	1.7 x 10^7
AsPh$_3$	6.5 x 10^4
P(n-Bu)$_3$	1.0 x 10^9
P(i-Pr)$_3$	6.7 x 10^7
P(O-i-Pr)$_3$	3.1 x 10^7

The reactions of CO with the substituted radicals {•Mn(CO)$_3$(PR$_3$)$_2$} are second-order, with rate constant values of 42 and 0.32 M^{-1} s^{-1} (24°C in hexane)[47] for R = *n*-Bu and *i*-Bu, respectively. Poë and co-workers[48] have shown that substitution on {•Re(CO)$_5$} is a second-order process.

Trogler and co-workers[49] studied the reactions of {•Fe(CO)$_3$(PR$_3$)$_2$$^+$} radicals and found that CO substitution is a second-order process whose rate depends on the steric bulk of the entering group. The rate constants correlate with the pK$_a$ for a series of pyridine nucleophiles. The parameters for pyridine reacting with {•Fe(CO)$_3$(PPh$_3$)$_2$$^+$} (25°C in CH$_2Cl_2$) are typical: k = 13.6 M^{-1} s^{-1}, ΔH* = 9.8 kcal mol^{-1}, ΔS* = –21 cal mol^{-1} K^{-1}.

Theoretical aspects of the substitution on 17-electron systems have been discussed recently by Therien and Trogler,[50] including an analysis of the geometries of the radicals and the direction of nucleophilic attack. The prevalence of associative attack is consistent with the observations of Basolo and co-workers[5] on various 17-electron V(0) species.

5.2 INSERTION REACTIONS

5.2.a CO "Insertion"

The classic example of a CO insertion reaction is

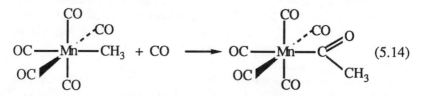

$$(5.14)$$

The CO insertion can be an important step in carbon–carbon bond-forming reactions that are catalyzed by organometallic complexes. At first sight, this appears to be an insertion of the entering CO into the Mn—CH$_3$ bond and the name insertion has continued to be used. However, phosphines and other nucleophiles (L) bring about an analogous transformation, as shown in Eq. (5.15):

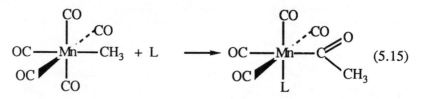

$$(5.15)$$

This could still be an insertion of a coordinated CO into the M—CH$_3$ bond, but it might also be migration of the CH$_3$ to a coordinated CO. The classic infrared study of Noack and Calderazzo[51] using ^{13}C-labeled CO showed that the reaction is actually -CH$_3$ migration. The original work has been confirmed by ^{13}C nmr studies by Flood et al.[52]

The basis of the mechanistic conclusion for this system relies on the product distribution from the reverse reaction (decarbonylation of CO), as shown in Scheme 5.5. If the mechanism is CH$_3$ migration, then the products should be 25 percent with the labeled CO trans to CH$_3$, 50 percent with the labeled CO cis to the CH$_3$ and 25 percent with no label, as shown in the diagram. If the reaction goes by insertion of a coordinated CO, then the products should be 75 percent with labeled CO in the cis position and 25 percent with no label. The results of both studies give a *product distribution consistent with CH$_3$ migration.*

Scheme 5.5

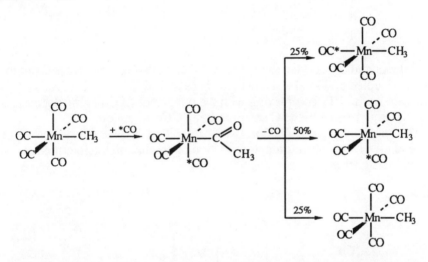

For many other metal carbonyls the mechanistic details are not known, and it is assumed that they are also proceeding by the methyl migration mechanism. The results of Wright and Baird[53] on the reaction of ^{13}CO with Fe(CO)$_2$(PMe$_3$)(I)(CH$_3$) are consistent with methyl migration, but the interpretation is complicated by I$^-$ dissociation.

The CO insertion in optically active (η^5-C$_5$H$_5$)Fe(CO)(PR$_3$)(CH$_3$) was studied by Flood and Campbell,[54] who expected the products to reflect which group migrates, as shown in Scheme 5.6. The observations are that in nitromethane and acetonitrile the products are consistent with methyl migration, while in dimethylsulfoxide, *N,N*-dimethylformamide, propylene carbonate, and hexamethylphosphoramide the products imply CO migration. However, the possible intervention of the η^2-acyl intermediate, shown in curled brackets in Scheme 5.6, makes the interpretation less than definitive with regard to the migrating group.

Scheme 5.6

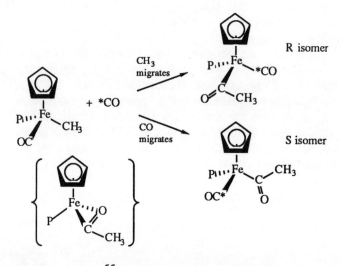

Brunner and co-workers[55] have shown that methyl migration occurs in the system studied by Flood if the reaction is catalyzed by BF_3. This catalysis is consistent with BF_3 complexing with the oxygen of CO, thereby decreasing the electron density on the C and promoting CH_3 migration.

The kinetics for these reactions with $Mn(CO)_5(CH_3)$ are consistent with a dissociative mechanism through formation of an unsaturated intermediate, as shown in Scheme 5.7.

Scheme 5.7

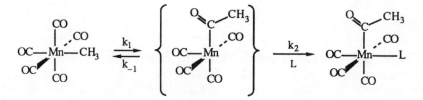

Then the pseudo-first-order rate constant ($[L] \gg [Mn]$) is given by

$$k_{exp} = \frac{k_1 k_2 [L]}{k_{-1} + k_2 [L]} \tag{5.16}$$

If L = CO, it is found that k_{exp} shows a direct dependence on [CO]. This is interpreted to mean that $k_2[CO] \ll k_{-1}$ because of the small CO concentration imposed by the low solubility of CO in most solvents. If L = pyridine,[56] then $k_{exp} = k_1$, apparently because $k_2[L] \gg k_{-1}$.

The reaction rate shows a substantial dependence on the nature of the solvent,[57] with faster reactions in more polar solvents. The solvent effect could be due either to better solvation of the polar intermediate or to direct coordination of the solvent. Even the rate law may change;[58] for CpMo(CO)$_3$(CH$_3$) reacting with PPh$_3$, the rate is first-order in [PPh$_3$] in benzene, but independent of [PPh$_3$] in tetrahydrofuran. Recently, Wax and Bergman[59] have studied this system in a series of methyl-substituted tetrahydrofuran solvents that were chosen because of their similar polarities. Their results are consistent with Scheme 5.8.

Scheme 5.8

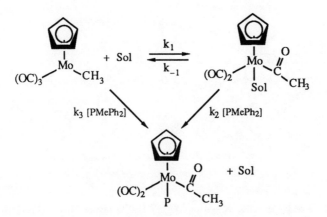

The pseudo-first-order rate constant is given by Eq. (5.17):

$$k_{exp} = \frac{k_1 [PMePh_2]}{\dfrac{k_{-1}}{k_2} + [PMePh_2]} + k_3 [PMePh_2] \qquad (5.17)$$

In THF and 3-MeTHF, the kinetics show saturation, and one can calculate k_{-1}/k_2; but in 2-MeTHF and 2,5-Me$_2$THF no saturation behavior is observed, indicating that k_{-1}/k_2 is much smaller. It is assumed that the latter solvents are much more weakly coordinated, so that k_2 is larger. The kinetic results are summarized in Table 5.6. Studies in mixed THF and 2,5-Me$_2$THF also show that the k_1 path is first-order in THF, as expected. The k_1 and k_2 steps appear to be associative because of the first-order dependence on the entering group. This might be caused by slippage of the η^5-C$_5$H$_5$ to an η^3-C$_5$H$_5$ form to allow for coordination of the entering group. It may be noted that the k_1 values show the trend expected for steric inhibition of the entering group with associative activation. The k_3 values are relatively constant, as expected if the general solvent effects are not large. The earlier work of Butler et al.[58] found the rate with PPh$_3$ in THF to be independent of [PPh$_3$], which implies that k_3 and k_2 are much smaller with this phosphine. This would be consistent with a steric effect of the entering group.

Table 5.6. Methyl Migration Rate Constants (59.9°C) for $CpMo(CO)_3(CH_3)$ in Various Solvents

Solvent	$10^4 \times k_1$ (s^{-1})	k_{-1}/k_2 (M)	$10^4 \times k_3$ $(M^{-1} s^{-1})$
	7.78	0.0104	1.73
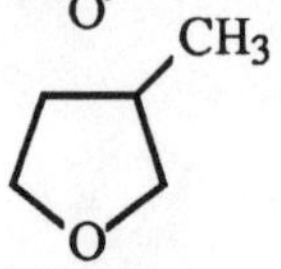	6.46	0.0082	1.86
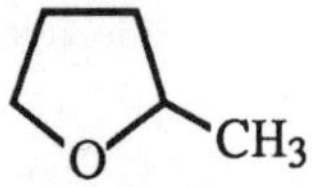	1.48		1.95
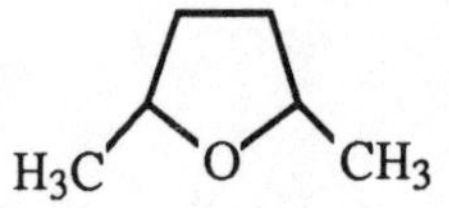	0.23		1.67

Insertion reactions on $R—Fe(CO)_4^-$ have been studied kinetically by Collman et al.[60] This work shows the importance of ion pairs in these reactions, a factor that must be remembered especially when charged species are involved. The system can be described by Scheme 5.9, which includes ion pairs and ion triplets with the reactant and product and where $Z^+ = Na^+$, Li^+, or $(Ph_3P)_2N^+$.

Scheme 5.9

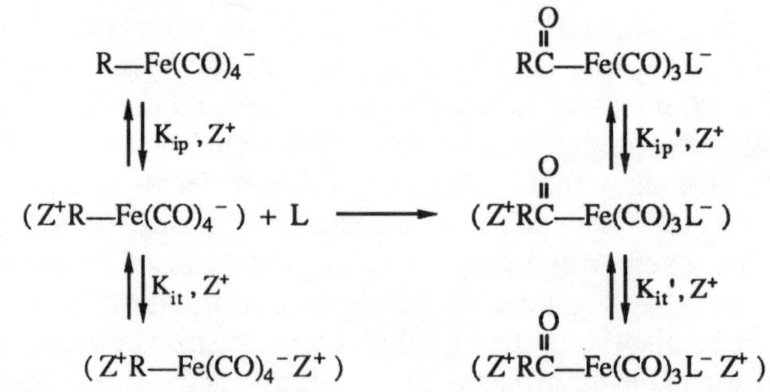

For Na^+ in THF, $K_{ip} = 1.1 \times 10^4$ M, $K_{ip}' = 2 \times 10^6$ M, $K_{it} = 1.7 \times 10^2$ M, and $K_{it}' = 1 \times 10^2$ M. The rate is first-order in [L] and in the reactant ion pair concentration. In this system, an increase in the solvent polarity reduces the rate, presumably because of less ion pair formation. The ion pair may be the more reactive species because of cation interaction with the CO ligands (Fe—CO···Na$^+$) which will favor methyl migration.

5.2.b   Sulfur Dioxide Insertion

Sulfur dioxide insertion has been found to proceed via the O-bonded sulfinate intermediate, which rearranges to the S-bonded product,[61] as shown in Scheme 5.10.

Scheme 5.10

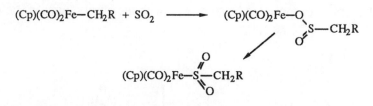

It has been observed[62] that this reaction proceeds with inversion at the alkyl carbon bonded to M, and the mechanism is believed to involve electrophilic attack of SO_2 at this C followed by rearrangement to the O-bonded isomer. An analogous pathway has been found[63] for SO_2 insertion into $(W(CO)_5Y(CH_3)_3)^-$, where Y= Si or Sn.

5.2.c   Carbon Dioxide Insertion

The type of reaction represented by carbon dioxide insertion is given in Eq. (5.18) and is another potentially important C—C bond forming process:

$$M—R + CO_2 \longrightarrow \begin{matrix} M—O \\ \diagdown \\ {}_O{\diagup}^{\diagup}C—R \end{matrix} \qquad (5.18)$$

The kinetics and stereochemistry have been investigated recently by Darensbourg and co-workers.[64] The rate is increased by more electron-rich metal complexes and shows much less sensitivity to the nature of the carbon center than the CO insertions. The reactions of $W(CO)_4(L)(R)^-$ (L = CO, phosphine, phosphite; R = Me, Et, Ph) have been studied in THF. The kinetics for $W(CO)_5CH_3^-$ show a first-order dependence on $[CO_2]$ and metal complex. The rate is increased by addition of Na$^+$, presumably due to ion pair formation. The kinetic work used $(Ph_3P)_2N^+$ as the counter ion, which should at least minimize ion pair effects because of the size and charge delocalization in this cation. For $W(CO)_4(P(OMe)_3)(CH_3)^-$, the activation parameters are $\Delta H^* = 42.7$ kJ mol^{-1} and $\Delta S^* = -181$ J mol^{-1} K^{-1}.

The stereochemistry at the carbon bonded to W was investigated using the threo-ligand isomer of $(OC)_5W$—CHD–CHDPh$^-$ and was found to proceed with retention of stereochemistry at the α carbon, as shown in Eq. (5.19).

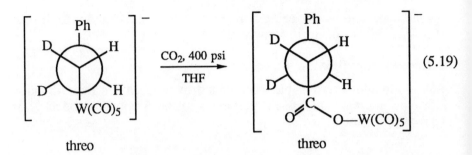

$$(5.19)$$

threo   threo

The mechanism proposed to be consistent with the observations is shown in Scheme 5.11.

Scheme 5.11

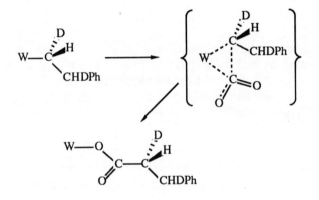

5.3 OXIDATIVE ADDITION REACTIONS

5.3.a General Considerations and Mechanisms

Oxidative addition reactions usually involve a coordinatively unsaturated 16-electron metal complex or a five-coordinate 18-electron species, and take the general form

$$(L)_n M + XY \longrightarrow (L)_n M(X)(Y) \qquad (5.20)$$

If the X and Y ligands in the product are considered to be formally −1, then the metal center has increased its formal oxidation state by +2, and this is the origin of the name *oxidative addition*. The reverse reaction is called *reductive elimination*. Some specific oxidative addition reactions are given in Eq. (5.21). The stereochemistry of the products can be controlled by subsequent isomerization reactions and is not always indicative of the immediate oxidative addition product.

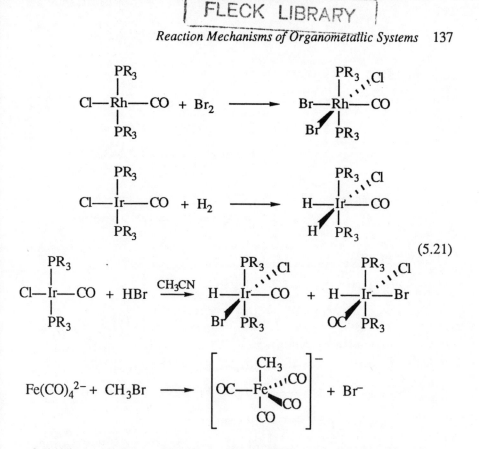

$$(5.21)$$

Oxidative addition reactions can also produce less obvious products, as shown in the following examples:

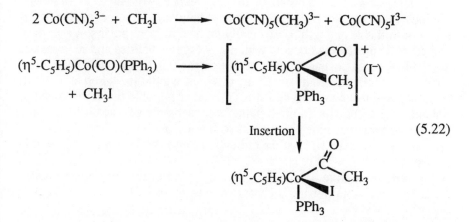

$$(5.22)$$

The rate law for these reactions is usually first-order in the metal complex and in XY concentrations. The reaction mechanism depends on the nature of XY and can be broken into three generally recognized categories, as shown in Scheme 5.12.

Scheme 5.12

Ionic

$$XY \rightleftharpoons X^+ + Y^-$$

$$(L)_nM + Y^- \longrightarrow \{(L)_nM(Y)\}^- \xrightarrow{X^+} (L)_nM(Y)(X)$$

Nucleophilic Attack (S_N2)

$$(L)_nM + XY \longrightarrow \{(L)_nM\text{---}X\text{--}Y\} \longrightarrow (L)_nM(Y)(X)$$

Free Radical (Single Electron Transfer, SET)

$$(L)_nM + XY \rightleftharpoons \{(L)_nM\bullet\}^+ + \{\bullet XY\}^-$$

$$\{\bullet XY\}^- \longrightarrow \{\bullet X\} + Y^-$$

Possible radical termination steps

$$\{(L)_nM\bullet\}^+ + \{\bullet X\} \longrightarrow \{(L)_nM\text{---}X\}^+ \xrightarrow{Y^-} (L)_nM(Y)(X)$$

$$\{(L)_nM\bullet\}^+ + Y^- \longrightarrow \{(L)_nM\text{---}Y\} \xrightarrow{\bullet X} (L)_nM(Y)(X)$$

$$(L)_nM + \{\bullet X\} \longrightarrow \{(L)_nM\bullet\}^+ + X^-$$

$$\{(L)_nM\bullet\}^+ + XY \longrightarrow \{(L)_nM\text{---}X\}^+ + \{\bullet Y\}$$

For species that are known to ionize, such as HI and HBr (and less obvious examples, such as $SnCl_4$ and acyl halides), the ionic mechanism is the most probable and will be especially favored in more polar solvents. The nucleophilic mechanism occurs with many organic halides and requires the availability of an unshared electron pair on the $(L)_nM$ species. The radical electron-transfer mechanism is an obvious candidate when XY is an oxidizing agent such as Cl_2 and Br_2. The other possible reactions, noted in Scheme 5.12 for the radical path, take account of the fact that the concentration conditions are such as to favor reaction of the radicals with reagents rather than with other radicals, unless the latter are reasonably persistent.

A fourth alternative, which is not widely encountered in organometallic systems, is the atom transfer mechanism in Scheme 5.13.

Scheme 5.13

Atom Transfer

$$(L)_nM + XY \longrightarrow (L)_nM\text{---}X + \bullet Y$$

$$(L)_nM + \bullet Y \longrightarrow (L)_nM\text{---}Y$$

If $(L)_nM$—X and $(L)_nM$—Y are stable products, the implication is that $(L)_nM$ is an odd-electron (e.g., 17-electron) organometallic complex. This mechanism is observed for $Co(CN)_5^{3-}$ and bis(dimethylglyoxime)cobalt(II) reacting with organic halides, but only the products are true organometallic complexes.

5.3.b Oxidative Addition of H_2

In view of the fact that the H—H bond energy is 105 kcal mol^{-1}, this reaction is unexpectedly facile with 16-electron complexes. The ΔH^* values are typically in the range of 5 to 10 kcal mol^{-1} and the ΔS^* values are in the range of -20 to -50 cal mol^{-1} K^{-1}. The reaction has always been observed to give the *cis*-dihydride product. The rate is increased by more electron-donating ligands on the metal. Some typical data are given in Table 5.7. The reaction shows a relatively modest deuterium isotope effect of ~ 1.2, which indicates that the H—H bond is largely intact in the transition state.

There has been a great deal of interest in this reaction because of its importance in catalytic hydrogenation reactions. Crabtree and co-workers[65] have discussed the stereochemistry of addition to d^8 complexes. Recently, complexes containing η^2-H_2 have been isolated[66] and characterized. One such complex is $W(CO)_3(PR_3)_2(H_2)$, which has a basic pentagonal bipyramid structure with an H—H bond length of 0.82 Å and a W—H length of 1.89 Å. The theoretical aspects of the bonding and stability have been discussed,[67,68] with the conclusion that the stability is strongly dependent on the π electron acceptor properties of the other ligands on the metal, especially CO in the preceding example. This is consistent with the observation that the isoelectronic $W(PR_3)_5$ complexes react with H_2 to give

Table 5.7. Rate Constants (30°C) and Activation Parameters for the Oxidative Addition of H_2 to $Ir(CO)(X)(PR_3)_2$ in Benzene

X	R	k (M^{-1} s^{-1})	ΔH^* (kcal mol^{-1})	ΔS^* (cal mol^{-1} K^{-1})
Cl	Ph	0.93 [a]	10.8	-23
Br	Ph	14.3 [a]	12.0	-14
I	Ph	>100 [a]		
Cl	p-OCH$_3$Ph	0.66 [b]	6.0	-39
Cl	p-CH$_3$Ph	0.53 [b]	4.3	-45
Cl	p-ClPh	0.16 [b]	9.8	-28
Cl	p-FPh	0.25 [b]	11.6	-22

[a] Halpern, J.; Chock, P. B. *J. Am. Chem. Soc.* **1966**, *88*, 3511.

[b] Ugo, R.; Pasini, A.; Fusi, A.; Cenini, S. *J. Am. Chem. Soc.* **1972**, *94*, 7364.

the classical dihydride product. It should be noted that dihydrides, $M(P(OR)_3)_5(H)_2$ (M = Cr and W),[69,70] have been found to be quite fluxional.

The reaction of H_2 with $Rh(Cl)(PPh_3)_3$ (Wilkinson's complex) is of special significance because of the applications of this complex as a catalyst. The system has been studied by Halpern and Wong[71] using stopped-flow spectrophotometry. The results show that even for 16-electron complexes the reaction may not be as straightforward as one might expect. The dependence of the rate on $[H_2]$ and $[PPh_3]$ is consistent with the mechanism in Scheme 5.14.

Scheme 5.14

$$Rh(Cl)(PPh_3)_3 + H_2 \xrightarrow{k_1} Rh(Cl)(PPh_3)_3(H)_2$$

$$Rh(Cl)(PPh_3)_3 \underset{k_{-2}}{\overset{k_2}{\rightleftharpoons}} Rh(Cl)(PPh_3)_2 + PPh_3$$

$$Rh(Cl)(PPh_3)_2 + H_2 \xrightarrow{k_3} Rh(Cl)(PPh_3)_2(H)_2$$

$$Rh(Cl)(PPh_3)_2(H)_2 + PPh_3 \xrightarrow{Fast} Rh(Cl)(PPh_3)_3(H)_2$$

The rate law for this mechanism is given by Eq. (5.23):

$$\text{Rate} = \left(k_1 + \frac{k_2 k_3}{k_{-2}[PPh_3] + k_3[H_2]} \right) [H_2] [Rh(Cl)(PPh_3)_3] \qquad (5.23)$$

The variation of the rate with $[PPh_3]$ and $[H_2]$ gives $k_1 = 4.8$ M^{-1} s^{-1}, $k_2 = 0.71$ s^{-1}, $k_{-2}/k_3 = 1.1$ (25°C in benzene). Note that at low $[PPh_3]$, the pseudo-first-order rate constant $k_{exp} = k_1[H_2] + k_2$, and for typical H_2 concentrations of $\sim 2 \times 10^{-3}$ M (at 1 atm), the k_2 path is dominant.

5.3.c   Oxidative Addition of Organic Halides

The available evidence indicates that oxidative addition of organic halides quite often proceeds by nucleophilic attack of the metal center on the halogen-bearing carbon, as shown in Scheme 5.15. In several examples, it has been shown that the reaction proceeds with inversion of configuration at C as predicted by the mechanism. The general example implies that addition of X^- is a subsequent process and it is usually too fast to be observed. The fact that these reactions are not very stereoselective is consistent with this. In the example in Scheme 5.15, the alkyl group and Cl^- are in a trans configuration in the product, and this would be difficult to rationalize if the addition were a concerted process.

Scheme 5.15

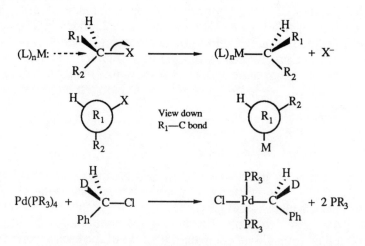

The rates of these reactions are first-order in each reactant, and some typical kinetic data are given in Table 5.8. The rate is much more affected by the ligand environment than is H_2 addition and increases with increasing polarity of the solvent. Decreasing the ligand basicity or increasing the steric bulk tends to slow down the rate of addition.[72] The order of reactivity of various halides is generally as follows:

$$CH_3 > CH_2CH_3 > CH(R)_2 > \text{cyclohexyl} > \text{adamantyl}$$

The S_N2 mechanism usually is observed with methyl, benzyl, and allyl halides, and α haloethers. But other saturated alkyl halides, vinyl and aryl

Table 5.8. Rate Constants[a] (25°C) and Activation Parameters for the Reaction of $CH_3I + Ir(Cl)(CO)(PR_3)_2$ in Benzene

R	k $(M^{-1} s^{-1})$	ΔH^* (kcal mol^{-1})	ΔS^* (cal mol^{-1} K^{-1})
Ph	3.3×10^{-3}	7.0	−47
Et(Ph)$_2$	1.2×10^{-2}	9.8	−34
(p-CH$_3$Ph)	3.3×10^{-2}	13.8	−20
(p-FPh)	1.5×10^{-4}	17.0	−20
(p-ClPh)	3.7×10^{-5}	14.9	−28

[a] Chock, P. B.; Halpern, J. *J. Am. Chem. Soc.* **1966**, *88*, 3511; Ugo, R.; Pasini, A.; Fusi, A.; Cenini, S. *J. Am. Chem. Soc.* **1972**, *94*, 7364.

halides, and α haloesters show characteristics of a radical pathway with $Ir(CO)(Cl)(PMe_3)_2$.[73] Direct epr evidence has been obtained for radicals in the reaction of $(\eta^5-C_5H_5)_2Zr(PMePh_2)$ with butyl chlorides.[74] The reaction of a gold(I) dimer, $(Au(H_2C)_2PPh_2)_2$, has been suggested[75] to proceed by an SET mechanism, based on the parallel between the rates and the reducibility of the organic halide. The reaction of alkyl halides with $Pt(PPh_3)_2$ is proposed to proceed by a halogen atom abstraction mechanism.[76]

5.4 REACTIONS OF ALKENES

Alkenes and alkynes are capable of normal substitution reactions like other nucleophiles but this is rarely an associative process with 18-electron systems because alkenes and alkynes are poor nucleophiles. Such systems require prior dissociation of a ligand to allow coordination of the alkene, which can then form strong complexes because of the back π bonding from the metal to the π^* orbital of the C—C multiple bond. The bonding π electrons have relatively low nucleophilicity. However, in coordinatively unsaturated systems, prior dissociation is not a problem. The expected initial reactions are shown in Scheme 5.16.

Scheme 5.16

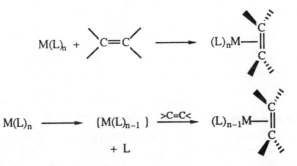

These reactions may be followed by rearrangement with H-atom migration to a σ-bonded form, as shown in Eq. (5.24):

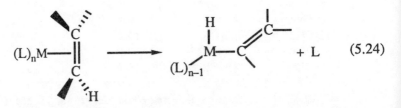

$$\tag{5.24}$$

The latter process would seem to be a straightforward rearrangement subsequent to π-complex formation, but recent work by Stoutland and

Bergman[77] indicates that the process is more complex. The thermal decomposition of $(\eta^5\text{-}C_5Me_5)Ir(PMe_3)(C_6H_{11})(H)$ yields C_6H_{12} and the unsaturated reactive species $(\eta^5\text{-}C_5Me_5)Ir(PMe_3)$. If the thermolysis is done in the presence of ethylene, the kinetic products are shown in Scheme 5.17.

Scheme 5.17

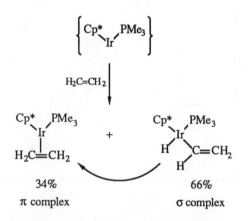

The σ complex will isomerize to the thermodynamically more stable π complex ($\Delta H^* = 34.6$ kcal mol^{-1}, $\Delta S^* = 2.6$ cal mol^{-1} K^{-1}), but the latter is the minor kinetic product. The implication is that the two products are forming directly from two slightly different transition states. Stoutland and Bergman have suggested that possible transition states might be the species shown in Figure 5.1, which involve initial bonding between the metal and hydrogen substituents on the ethylene. A perplexing feature of the observations is that the same isomer product distribution is obtained from $H_2C=CD_2$ and *cis*-(HD)C=C(HD). It was expected that an isotope effect acting on the proposed transition states would cause a change in product distribution for the deuterated reactants. A combination of calorimetric measurements and bond energy estimates has allowed construction of the reaction coordinate diagram in Figure 5.2. This diagram is different from the normal one in that the reaction starts at the high-energy intermediate, that

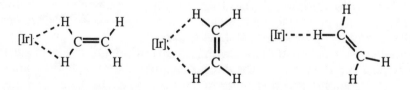

Figure 5.1. Possible transition states for the addition of ethylene to coordinatively unsaturated iridium.

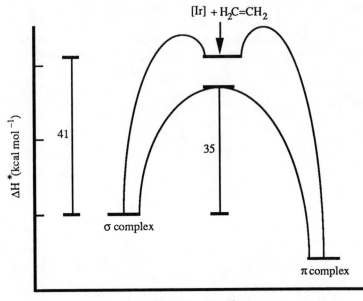

Figure 5.2. Reaction coordinate diagram for the addition of ethylene to coordinatively unsaturated iridium.

will then react to give the two products in amounts that depend on the relative heights of the energy barriers to the left and right. It should be noted that the difference in energy barriers must be less than 1 kcal mol^{-1} in order to explain the approximately two-to-one product distribution. The difference is greatly exaggerated in Figure 5.2, in order to show that the barrier leading to the π complex must be higher to account for the product distribution.

Baker and Field[78] have observed rather similar behavior with $((H_2C)_2(R)_4P_2)_2Fe(H)_2$, which after photolysis in the presence of ethylene gives mainly the cis σ complex. The latter isomerizes to trans and eventually yields the more stable π complex. Graham and co-workers[79] have observed that the trifluoromethylpyrazolylborate (HBPf$_3$) π complex of iridium $(\eta^2\text{-HBPf}_3)Ir(CO)(^2\eta\text{-}C_2H_4)$ converts to the more stable σ complex $(\eta^2\text{-HBPf}_3)Ir(CO)(H)(\text{-}C_2H_3)$ at 100°C, while the σ complex of rhodium converts to the more stable π complex at 25°C. Clearly, there is a delicate balance between the stabilities of these species.

An unusual reaction of olefins with $(\eta^5\text{-}C_5H_5)Co(NO)_2$ has been studied by Becker and Bergman.[80] The direct substitution reaction is first-order in the metal complex and first-order in olefin. The exchange of one olefin adduct with another also was studied. These reactions are shown in Scheme 5.18.

Scheme 5.18

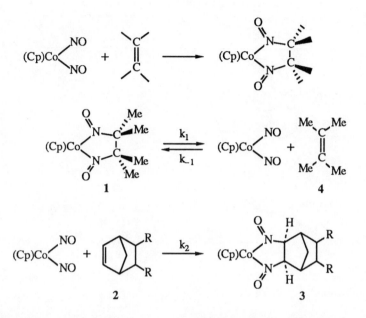

The rate law for the exchange reaction is consistent with the preceding mechanism if a steady-state condition is applied to $(Cp)Co(NO)_2$ to yield Eq. (5.25):

$$k_{exp} = \frac{k_1 [1][2]}{\left(\dfrac{k_{-1}}{k_2}\right)[4]+[2]} \tag{5.25}$$

The kinetic results (75°C in toluene) give $k_1 = 4.3 \times 10^{-4}$ s^{-1} ($E_a = 29.4$ kcal mol^{-1}) and $k_2/k_{-1} = 115$ for R = CO_2Me.

For the direct reaction of olefin with $(Cp)Co(NO)_2$, the second-order rate constant depends on the olefin, as indicated by some of the data in Table 5.9. The exchange of one olefin for another is stereospecific, with no isomerization of the olefin. This observation indicates that the exchange does not proceed through an intermediate with a C—C single bond, or at least that such an intermediate is not persistent enough to allow rotation about the C—C bond, as indicated in Scheme 5.19. Becker and Bergman favor a concerted mechanism (no intermediate) for the addition and dissociation reactions.

Since $(Cp)Co(NO)_2$ can achieve an 18-electron configuration by considering one NO as a three-electron donor (NO^+) and the other as a one-electron NO^-, this reaction could be considered as a formal analogue of 1,3-dipolar cycloadditions in organic chemistry. It is also possible for the NO^+

Table 5.9. Rate Constants (20°C) for the Reaction of Olefins with (Cp)Co(NO)₂ in Cyclohexane

Olefin	k_2 (M^{-1} s^{-1})
	130
	3.9
	0.84
	0.25

Scheme 5.19

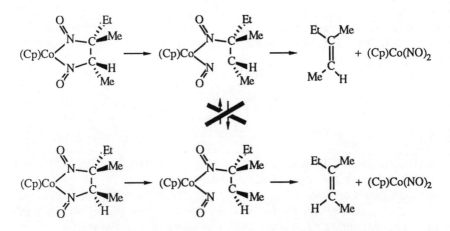

to switch to an NO⁻ in the transition state, thereby leaving the Co unsaturated and able to form a transient π complex before rearranging to the observed product. Slippage of the C₅ ring is another possibility, but the rate only changes by a factor of about three when the C₅ is changed from C₅H₅ to C₅H₄CO₂Me to C₅(Me)₅.

5.5 CATALYTIC HYDROGENATION OF ALKENES

The addition of H₂ to a >C=C< system is thermodynamically favorable but generally difficult to achieve. In the laboratory, chemists use catalysts such as platinum black and Rainey nickel to make the reaction proceed at a reasonable rate. The kinetic barrier for this reaction can be understood in terms of simple orbital symmetry diagrams. The basic principle is that in the activated state the electrons must flow in such a way as to make and break the appropriate bonds. In the case of hydrogenation, we want to break the H—H σ and C—C π bonds and make two C—H bonds. The electrons must flow from an occupied orbital on one molecule to an unoccupied orbital on the other, that is, from the highest occupied orbital (HOMO) on one species to the lowest unoccupied orbital (LUMO) on the other.

For H_2, the HOMO is the σ-bonding orbital and the LUMO is the corresponding σ-antibonding orbital. For an alkene, the HOMO is the π-bonding orbital and the LUMO is the corresponding π-antibonding orbital. To stabilize the activated complex, the appropriate HOMO and LUMO must produce good overlap; that is, the signs of the radial parts of the wave functions should be the same in the overlap region. The first diagram in Figure 5.3 shows that the overlap is correct for the two HOMO orbitals in this system, but there can be no useful electron flow between two occupied orbitals. The second and third diagrams show that the HOMO–LUMO combinations do not produce net overlap; thus, the transition state will not be stabilized. Such a reaction is said to be *symmetry forbidden*. A catalyst must somehow overcome this symmetry restriction.

5.5.a   General Mechanisms

There are two generally recognized routes by which an organometallic complex can catalyze the hydrogenation of alkenes. They are referred to as the *olefin route* and the *hydride route* and are shown in Scheme 5.20. Both start from a coordinatively unsaturated (16-electron) metal complex. The first step in the olefin route is addition of the olefin to form a π complex, whereas the first step in the hydride route is oxidative addition of H_2.

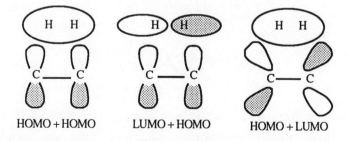

HOMO + HOMO	LUMO + HOMO	HOMO + LUMO

Figure 5.3. The highest occupied molecular orbitals of H_2 and CH_2CH_2, and their combinations with the lowest unoccupied molecular orbitals in each molecule. The shaded lobes of the orbitals have the same sign as the radial part of the wave function.

Scheme 5.20

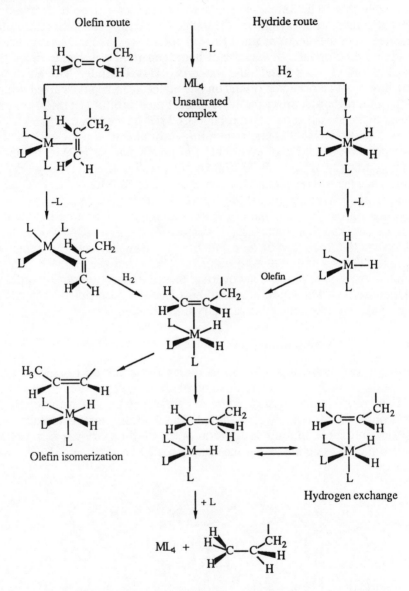

5.5.b Hydrogenation by Wilkinson's Catalyst: ClRh(PPh₃)₃

The catalytic properties of ClRh(PPh$_3$)$_3$ were first reported by Wilkinson and co-workers.[81] The nature of the species present in hydrocarbon solvents was the subject of controversy until the work of Arai and Halpern,[82] which indicated some PPh$_3$ dissociation, shown by

$$ClRh(PPh_3)_3 \rightleftharpoons ClRh(PPh_3)_2 + PPh_3 \qquad (5.26)$$

However, Tolman and co-workers[83] found that phosphine liberation was accompanied by dimer formation, which has K = 2.4×10^4 M in benzene at 25°C. The reaction and suggested structure for the dimeric product are given by

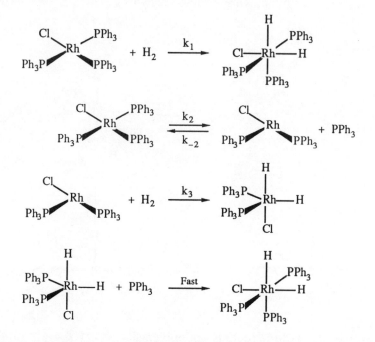

$$2\ ClRh(PPh_3)_3 \rightleftharpoons \underset{Ph_3P}{\overset{Ph_3P}{\diagdown}} Rh \overset{Cl}{\underset{Cl}{\diagdown\diagup}} Rh \underset{PPh_3}{\overset{PPh_3}{\diagup}} + 2\ PPh_3 \qquad (5.27)$$

Halpern and Wong[84] acknowledged the correctness of Tolman's interpretation and studied the kinetics of the hydrogenation of ClRh(PPh₃)₃ in the presence of excess PPh₃ to suppress dimer formation. With [H₂] and [PPh₃] >> [Rh] at 25°C in benzene, the kinetics indicate that the reaction proceeds by parallel paths involving oxidative addition to Wilkinson's catalyst and to the species with one phosphine dissociated as shown in Scheme 5.21.

Scheme 5.21

The rate law for Scheme 5.21 gives the following expression for the pseudo-first-order rate constant, if a steady state is assumed for ClRh(PPh₃)₂:

$$k_{exp} = k_1 + \frac{k_2}{\left(\dfrac{k_{-2}}{k_3}\right)[PPh_3] + [H_2]} \qquad (5.28)$$

The kinetic results give $k_1 = 4.8$ M^{-1} s^{-1}, $k_2 = 0.71$ s^{-1} and $k_{-2}/k_3 = 1.1$.

Separate studies on the dimer gave a rate that is first-order in [dimer] and [H$_2$] and independent of [PPh$_3$], with $k = 5.4$ M^{-1} s^{-1}. Tolman et al.[85] showed that the dimer reacts with H$_2$ and with ethylene to give the following dihydride and ethylene complexes, but it does not react with cyclohexene.

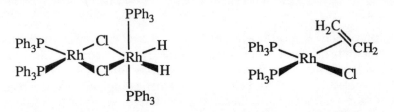

Halpern et al.[86] found that the hydrogenation of cyclohexene by the rhodium dihydride has kinetics consistent with Scheme 5.22, with $K_5 = 3.4$ x 10^{-4} and $k_6 = 0.2$ s^{-1} at 25°C in benzene. These examples are considered to be consistent with the hydride route for hydrogenation with Wilkinson's catalyst.

Scheme 5.22

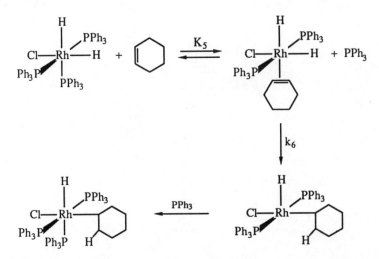

However, the hydride route is not universally observed even for closely related catalysts. Halpern and co-workers[87] studied the diphos system and found that the kinetics were consistent with the reaction sequence shown in Scheme 5.23 . The reaction follows the "olefin route," and the kinetics give

$K_4 = 1.6$ M^{-1} and $k_6 = 0.18$ atm^{-1} s^{-1} at 25°C. Halpern has rationalized the difference caused by using a chelating phosphine as due to the instability of dihydride that forms by oxidative addition of H$_2$. The trans configuration of the hydride and phosphine ligands is thought to be unstable because of the trans effect of the phosphine.

Scheme 5.23

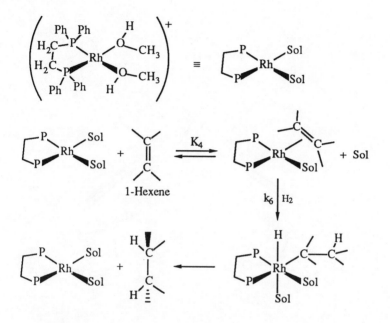

5.5.c Asymmetric Hydrogenation

An important commercial application of these types of catalysts is in the production of L-dopa for the treatment of Parkinson's disease. The key to this application is the stereoselectivity shown by the phosphine chelate 2S,3S bis(diphenylphosphino)butane, called chiraphos. Halpern and co-workers[88] have shown that the system is unusual in that the most stable olefin adduct does not lead to the major or desired product. The system also follows the "olefin route" like the chelate in Scheme 5.23. The kinetic components of various steps have been studied recently by Landis and Halpern,[89] who conclude that the minor species generates the major portion of the product because of the greater reactivity of the intermediate with H$_2$. The mechanistic steps and stereochemical results are described in Scheme 5.24.

The field of asymmetric hydrogenation has been reviewed recently.[90] There are a number of examples of such reactions in which the mechanistic details are uncertain. Recently Takahashi et al.[91] have reported highly selective hydrogenation of 3-(aryloxy)-2-oxo-1-propylamine derivatives with a chiral phosphine of Rh(I).   Chan and Osborn[92] found an Ir(III) diphosphine-monohydride which gives enantioselective hydrogenation of imines.   In a heterogeneous system, Garland and Blaser[93] have reported good enantioselectivity for the hydrogenation of ethyl pyruvate by Pt on Al_2O_3 modified with 10,11-dihydrocinchonidine.

Scheme 5.24

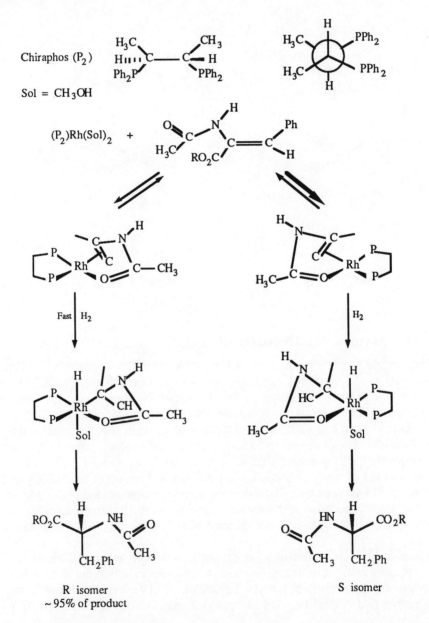

R isomer
~ 95% of product

S isomer

5.5.d Carbon–Hydrogen Bond Activation

Carbon–hydrogen bond activation is formally the oxidative addition of a hydrocarbon to a metal complex, as shown in reaction (5.29). It is a potentially important reaction because it represents the initial step in a possible route to functionalize hydrocarbons.

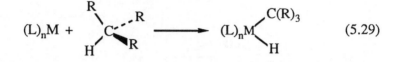

$$(5.29)$$

Until the discovery of this process in 1982,[94,95] it was thought to be difficult at best and perhaps impossible under moderate conditions because of the low reactivity of hydrocarbons and the high strength (95–100 kcal mol⁻¹) of the C—H bond. However, the first observations of this type of reaction indicated that it could occur under relatively mild conditions, as shown in the following examples from the first studies:

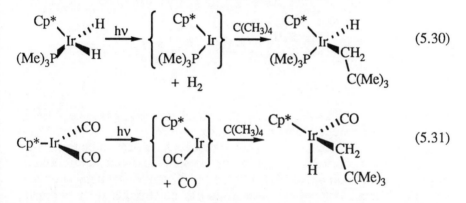

In both cases, a highly reactive unsaturated species is assumed to be generated by photolysis and reacts with the neopentane solvent as shown. Cyclohexane reacted similarly in both cases. Recently, Ghosh and Graham[96] have found that the tris(dimethylpyrazolyl)borato complex (HBPz*₃)Rh(CO)₂ can be photolyzed under milder conditions with elimination of CO and oxidative addition of hydrocarbons. In this system, the cyclohexyl hydride complex exchanges with methane under thermal conditions to give the methyl hydride complex, as shown in Eq. (5.32):

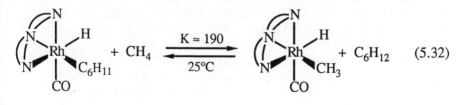

The first generation of functionalized products from such reactions has been demonstrated by Ghosh and Graham,[97] using (HBPz*$_3$)Rh(CO)(C$_2$H$_4$), as shown in Scheme 5.25.

Scheme 5.25

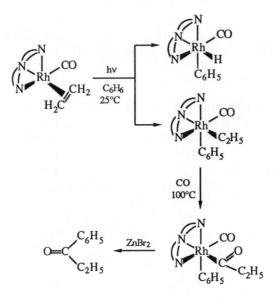

Since the original discovery, many more metal complexes have been found to undergo oxidative addition of hydrocarbons.[98-101]

The energetics of the C—H activation process have been of concern since its discovery. Calorimetric, equilibrium constant, and kinetic information have been used to obtain estimates of the M—C and M—H bond strengths. In general, these bonds have been found to be stronger than originally anticipated. For the simple oxidative addition of a hydrocarbon, the enthalpy change can be estimated as

$$\Delta H_{rxn} = BE(C—H) - [BE(M—H) + BE(M—C)] \qquad (5.33)$$

It is commonly assumed that the reaction will proceed if it is exothermic (i.e., entropy effects are assumed to be small, so that $\Delta H_{rxn} \approx \Delta G_{rxn}$). Thus, the bond energies can be used to predict whether an oxidative addition of a hydrocarbon is probable. Some bond energies[102] for different metal centers are given in Table 5.10. The C—H bond strengths, in kcal mol^{-1}, are 96 for C$_6$H$_{12}$, 110 for C$_6$H$_6$, and 98 for CH$_4$, and the C—C bond strength is 83 in alkanes. From the data in Table 5.10, one can estimate that for the oxidative addition of methane to Cp$_2$Th, $\Delta H_{rxn} \approx 98 - (97.5 + 82)$ ≈ -82 kcal mol^{-1}, and the reaction should be quite exothermic. However, it is of greater interest and relevance to estimate the relative stabilities of the

Table 5.10. Bond Energies for M—X in L_nMX_2 Systems (kcal mol^{-1})

X	$(Me_3P)Cp^*IrX_2$	Cp_2WX_2	Cp_2ThX_2	$L_2Cl(CO)IrX_2$
H	74	73	97.5	60
Cl	90	83		71
Br	76	71.5		53
I	64	64		35
CH_3			82	35
C_6H_5	81		91	
C_6H_{11}	51		75	

reactants and products for overall reactions such as

$$(Me_3P)Cp^*Ir(H)_2 + C_6H_{12} \longrightarrow (Me_3P)Cp^*Ir(H)(C_6H_{11}) + H_2 \quad (5.34)$$

whose enthalpy change is given by

$$\Delta H_{rxn} = BE(C\text{—}H) + 2\,BE(Ir\text{—}H) - BE(Ir\text{—}H) - BE(Ir\text{—}C) - BE(H\text{—}H)$$
$$= 96 + (2 \times 74) - 74 - 51 - 104 = 15 \text{ kcal mol}^{-1} \quad (5.35)$$

A notable feature in Table 5.10 is the strength of the M—C_6H_5 bond. This implies that benzene can be expected to replace other oxidatively added hydrocarbons in an exchange reaction. Since the C—C bond in alkanes is weaker than the C—H bond, one might anticipate C—C bond activation, as in the following example:

$$(Cp)_2Th(H)_2 + H_3C\text{—}CH_3 \longrightarrow (Cp)_2Th\begin{smallmatrix} \diagup CH_3 \\ \diagdown CH_3 \end{smallmatrix} + H_2 \quad (5.36)$$

$$\Delta H_{rxn} = BE(C\text{—}C) + 2\,BE(Th\text{—}H) - 2\,BE(Th\text{—}C) - BE(H\text{—}H)$$
$$= 83 + (2 \times 97.5) - (2 \times 82) - 104 = 10 \text{ kcal mol}^{-1} \quad (5.37)$$

For comparison, the corresponding C—H oxidative addition gives

$$\Delta H_{rxn} = BE(C\text{—}H) + 2\,BE(Th\text{—}H) - BE(Th\text{—}H) - BE(Th\text{—}C) - BE(H\text{—}H)$$
$$= 98 + (2 \times 97.5) - 97.5 - 82 - 104 = 9.5 \text{ kcal mol}^{-1} \quad (5.38)$$

Thermodynamics allows no clear choice between these two possibilities and kinetic factors must be at work to favor C—H activation.

Mechanistic studies on these reactions have been limited because of the necessity of generating the unsaturated metal center photochemically and the need for an appropriate "inert" solvent also is a problem.

Marx and Lees[103] studied quantum yields (Φ) for the photochemical reaction of $CpIr(CO)_2$ with C_6H_6 in hexafluorobenzene at 20°C and found that Φ shows a saturation effect with increasing $[C_6H_6]$ and is independent of [CO]. They concluded that the rate-controlling process is not CO dissociation and suggest that it is η^5 to η^3 slippage, which then allows complexation of benzene prior to oxidative addition. Drolet and Lees[104] have proposed a similar slippage mechanism for the photolysis of $CpRh(CO)_2$ in the presence of PPh_3 in decalin at 10°C. Their conclusion is based on the observation that the quantum yield for CO replacement is first-order in $[PPh_3]$ and unaffected by 9×10^{-3} M CO.

Bergman and co-workers[105] have studied the reaction of $Cp*Rh(CO)_2$ with C_6H_{12} by infrared laser flash kinetics in inert gas solvents. Their conclusions are different from those of Lees and co-workers. At –31°C in xenon, irradiation gives a new species that Bergman and co-workers assign as $Cp*Rh(CO)$ or its solvated form $Cp*Rh(CO)(Xe)$. A similar but more reactive species is observed in krypton. When cyclohexane is added to the krypton solution, this species undergoes oxidative addition with a rate that shows saturation behavior described by

$$k_{exp} = \frac{\alpha\,[C_6H_{12}]}{[C_6H_{12}] + \beta} \tag{5.39}$$

If the mechanism involves rate-controlling dissociation of the solvent followed by cyclohexane addition, then α should be independent of the nature of the alkane. However, if C_6D_{12} is used, α decreases by a factor of about 20. Therefore, Bergman and co-workers have proposed the mechanism in Scheme 5.26.

Scheme 5.26

$$Cp*Rh(CO)(Kr) + C_6H_{12} \underset{}{\overset{K_{eq}}{\rightleftharpoons}} Cp*Rh(CO)(C_6H_{12}) + Kr$$

$$\Big\downarrow k_1$$

$$Cp*Rh(CO)(H)(C_6H_{11})$$

From the preceding expression for k_{exp}, $\alpha = k_1$ and $\beta = [Kr]/K_{eq}$. The nature of the cyclohexane precursor complex is uncertain.

In a related study, Perutz and co-workers[106] observed the time-resolved photochemistry of $CpRh(CO)_2$ and $(CpRh(CO))_2(\mu\text{-}CO)$ in the 300-nm region. The results are summarized in Scheme 5.27, where RH is a solvent such as C_6H_{12} or C_6H_6. The structure of $(CpRh(\mu\text{-}CO))_2$ was assigned as shown in the scheme on the basis of its polarized infrared spectrum. The unusual aspects of these observations are the high dimerization rate and the rapid formation of the hydride species of cyclohexane within 400 ns (the product with benzene could be η^2-C_6H_6).

Scheme 5.27

$$CpRh(CO)_2 \xrightarrow{h\nu} CpRh(CO) + CO$$

$$CpRh(CO) + RH \underset{k = 2.7 \times 10^3 \, s^{-1}}{\overset{k \geq 3 \times 10^5 \, M^{-1} s^{-1}}{\rightleftharpoons}} CpRh(CO)(H)(R)$$

$$CpRh(CO)(H)(R) + L \xrightarrow{k \approx 3 \times 10^8 M^{-1} s^{-1}} CpRh(CO)(L) + RH$$

$$CpRh(CO) + CpRh(CO)(H)(R) \xrightarrow{k \approx 10^{10} \, M^{-1} s^{-1}} [CpRh(\mu\text{-}CO)]_2 + RH$$

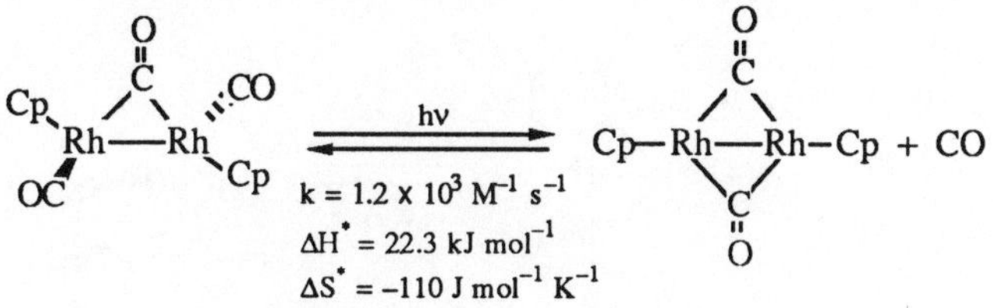

Studies of the reverse reaction, reductive elimination, have also been used to shed light on the overall reaction mechanism. Norton and co-workers[107] studied the reductive elimination of methane from $Cp_2W(H)(CH_3)$ and found that H exchange into the CH_3 group occurs more rapidly than CH_4 elimination. Such exchange has been observed in other systems, and some rate constants at 48°C in 10 percent CD_3CN/90 percent C_6D_6 are given in Scheme 5.28.

Scheme 5.28

$$Cp_2W(D)(CH_3) \underset{k = 5 \times 10^{-6} \, s^{-1}}{\overset{k = 2.5 \times 10^{-5} \, s^{-1}}{\rightleftharpoons}} Cp_2W(H)(CDH_2)$$

$$Cp_2W(H)(CH_3) \xrightarrow[\substack{\Delta H^\ast = 24.5 \, \text{kcal mol}^{-1} \\ \Delta S^\ast = -5.6 \, \text{cal mol}^{-1} K^{-1}}]{k = 8.4 \times 10^{-6} \, s^{-1}} \{Cp_2W\} + CH_4$$

Elimination has an inverse isotope effect (H/D) of 0.75 at 72.6°C. The authors propose that the exchange and reductive elimination proceed through a σ complex that may revert to reactant with exchange or give elimination, as shown in Scheme 5.29.

Scheme 5.29

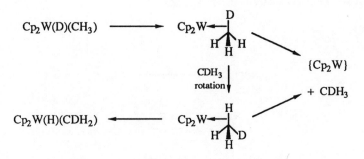

Bergman and co-workers[108] studied the exchange reaction in Scheme 5.30 and found that the kinetics are consistent with the mechanism shown.

Scheme 5.30

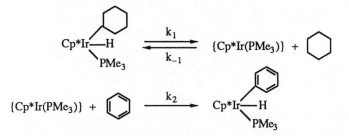

The pseudo-first-order rate constant is given by the following equation, and the reaction rate is unaffected by PMe_3:

$$k_{exp} = \frac{k_1 \, [C_6H_6]}{\left(\dfrac{k_{-1}}{k_2}\right)[C_6H_{12}] + [C_6H_6]} \qquad (5.40)$$

The reaction was confirmed to be intramolecular and not to involve H transfer to Cp^*, since $Ir(H)(C_6H_{11}) + C_6D_6$ yields only C_6H_{12} with no isotope scrambling. This seems to eliminate radical mechanisms involving Ir—C bond homolysis. Ring slippage is compatible with the rate law only if the oxidative addition of C_6H_{12} to the intermediate is competitive with the η^3 to η^5 change to give product; Bergman and co-workers believe that this is unlikely. There is an inverse isotope effect on k_1 $(H/D) \approx 0.7$, and the activation parameters for k_1 are $\Delta H^* = 35.6$ kcal mol^{-1} and $\Delta S^* = 10$ cal mol^{-1} K^{-1}. Isotope exchange was observed in $Ir(D)(C_6H_{11})$, with the α position of C_6H_{11} becoming deuterated. All these observations are similar to those of Norton and co-workers, and an analogous σ complex would explain the results.

5.6 HOMOGENEOUS CATALYSIS BY ORGANOMETALLIC COMPOUNDS

There are a number of processes that appear to be catalyzed by organometallic complexes. The reactions generally involve C—C bond formation and often involve CO and H_2. In the industrial application, the catalyst is often added as a metal salt, and it is assumed that this is transformed to active organometallic species under the reducing conditions of the process. In most cases, there may be several organometallic species present, and the nature of the catalytic mechanism is inferred from the known chemistry of simpler systems and the overall rate law.

5.6.a Hydroformylation or "Oxo" Reaction

The overall process is described by

$$RCH=CH_2 \xrightarrow[CO]{H_2} RCH_2CH_2-C\begin{smallmatrix}O\\H\end{smallmatrix} \xrightarrow{H_2} RCH_2CH_2CH_2OH \qquad (5.41)$$

In 1938, Roelen discovered that the reaction is catalyzed by cobalt salts, and more recently, rhodium catalysts have been developed. The main commercial product is *n*-butanol from propylene and the major problems are to avoid branched-chain alcohols and alkanes. The reaction was reviewed by Pruett.[109] With cobalt as the catalyst, the reaction is done at 100°C to 200°C and at 100 to 500 psi of H_2 + CO. Linear product is favored by lower CO pressures and by "modified cobalt catalysts" that contain phosphines in the reaction mixture. The rate law is given by

$$\text{Rate} = k\,[\text{Co}]\,[\text{Alkene}]\,[H_2]\,[\text{CO}]^{-1} \qquad (5.42)$$

It is accepted generally that the important cobalt species is $HCo(CO)_4$ formed by the following reaction:

$$Co_2(CO)_8 + H_2 \rightleftharpoons 2\,HCo(CO)_4 \qquad (5.43)$$

The mechanism in Scheme 5.31 was first proposed by Heck and Breslow[110] and is still thought to be essentially correct. The hydride shift in the third step followed by CO "insertion" are common steps proposed in many such processes. The first dissociation of CO is assumed to be the source of the $[\text{CO}]^{-1}$ term in the rate law.

Wilkinson and co-workers[111] discovered that rhodium catalysts are effective in this reaction. It is thought that the key intermediate may be $HRh(CO)_2(PPh_3)_2$, which may react by dissociative loss of PPh_3 or by direct formation of a π-olefin complex. The reaction could proceed by a hydride route, as proposed in some hydrogenation processes.

Scheme 5.31

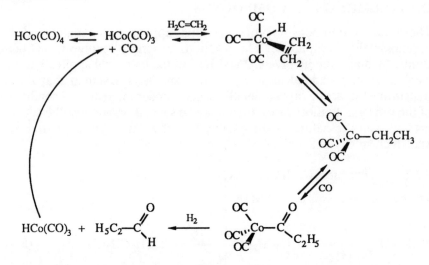

5.6.b   Reppe Synthesis

The Reppe synthesis, shown in Eq. (5.44), has a formal resemblance to hydroformylation, with H_2 replaced by H_2O:

$$RCH{=}CH_2 + 2\,H_2O + 3\,CO \longrightarrow R(CH_2)_2CH_2OH + 2\,CO_2 \quad (5.44)$$

The most effective catalysts for the reaction are iron species often derived from $Fe(CO)_5$. Pettit and co-workers[112] found that the catalyst is most effective at pH > 11. This has been rationalized by the known reactions shown in (5.45), and assuming that $H_2Fe(CO)_4$ is the active species that interacts with alkenes:

$$
\begin{array}{ccc}
Fe(CO)_5 & \xrightarrow{} & HFe(CO)_4^- \xrightarrow{H_2O} H_2Fe(CO)_4 \\
+\ OH^- & & +\ CO_2 \phantom{\xrightarrow{H_2O}} +\ OH^-
\end{array}
\quad (5.45)
$$

5.6.c   Fischer–Tropsch Reaction

The Fisher–Tropsch reaction involves the conversion of coal to hydrocarbons as follows:

$$C + H_2O \rightleftharpoons CO + H_2 \longrightarrow \text{Hydrocarbon products} \quad (5.46)$$

The process often is considered in conjunction with reaction (5.47), the water gas shift reaction, since this increases the H_2 content of the reactant mixture. The reaction is attractive because of the conversion of cheap materials, coal and water, to valuable hydrocarbons.

$$H_2O + CO \rightleftharpoons H_2 + CO_2 \qquad (5.47)$$

The process was developed by Fischer and Tropsch in 1923 using heterogeneous iron catalysts, and catalyst improvements have continued in a rather unsystematic way ever since. The reaction became uneconomical with the advent of cheap oil after World War II but came into vogue again in the late 1970s as oil prices rose. Only South Africa has operating plants using the process, with the industrial conditions being 200°C to 300°C and total pressure of ~25 atm.

It has been found by Muetterties and co-workers[113] and by Ford and co-workers[114] that polynuclear carbonyls such as $Os_3(CO)_{12}$, $Ru_3(CO)_{12}$, and $Ir_4(CO)_{12}$ are effective homogeneous catalysts. The apparent requirement for at least two metal centers in the catalyst has stimulated much work on dimetal species and metal cluster complexes. It has been suggested that a key intermediate is an η^2 CO in which the η^2 bonding serves to weaken the CO triple bond and make it susceptible to reduction. The general mechanism in Scheme 5.32, which was suggested by Masters,[115] also involves formation of a metal carbene ($M=CH_2$) as an important step in the process.

Scheme 5.32

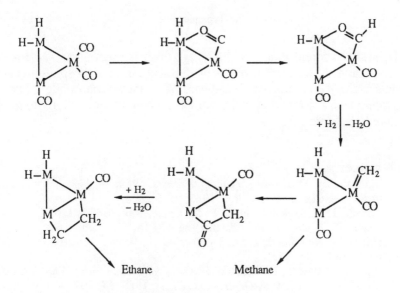

Ethane   Methane

5.6.d Ethylene–Butadiene Codimerization

The purpose of the ethylene–butadiene codimerization process is to produce *trans*-1,4-hexadiene from ethylene and butadiene as in reaction (5.48). The product is an important monomer in synthetic rubber production.

$$H_2C=CH_2 + H_2C=CHCH=CH_2 \longrightarrow H_2C=CHCH_2CH=CHCH_3 \quad (5.48)$$

The reaction is catalyzed in the presence of $RhCl_3$ in aqueous HCl and has been studied kinetically by Cramer.[116] It is proposed that under the reducing influence of the ethylene, a rhodium hydride species (Cl_2RhH ?) forms as the catalyst and reacts with ethylene to give an η^1-ethyl complex that may be converted to the η^2-ethyl complex by complexing with butadiene. This is followed by H transfer to give a π-crotyl complex and finally the desired product, as shown in Scheme 5.33.

Scheme 5.33

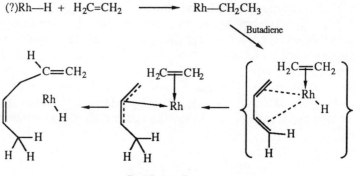

π-Crotyl complex

Tolman[117] found that $HNi(P(OEt)_3)_4$ also is a good catalyst for this reaction, and showed that it reacts by dissociation of a phosphite ligand to form a π-crotyl complex with butadiene. Then another phosphite is lost and the π-ethylene + π-crotyl complex forms and rearranges similarly to the Rh system in Scheme 5.33.

References
1.   Darensbourg, D. J. *Adv. Organomet. Chem.* **1982,** *21,* 113.
2.   Basolo, F. *Coord. Chem. Rev.* **1982,** *49,* 7.
3.   Howell, J. A. S.; Burkinshaw, P. M. *Chem. Rev.* **1983,** *83,* 557.
4.   Basolo, F. *J. Organomet. Chem.* **1990,** *383,* 579.
5.   Shi, Q.-Z.; Richmond, T. G.; Trogler, W. C.; Basolo, F. *J. Am. Chem. Soc.* **1984,** *106,* 71.
6.   Day, J. P.; Basolo, F.; Pearson, R. G. *J. Am. Chem. Soc.* **1968,** *90,* 6927.
7.   Noack, K.; Ruch, M. *J. Organomet. Chem.* **1969,** *17,* 309.
8.   Shen, J.-K.; Gao, Y.-C.; Shi, Q.-Z.; Basolo, F. *Inorg. Chem.* **1989,** *28,* 4304.
9.   Huq, R.; Pöe, A. J.; Chawla, S. *Inorg. Chim. Acta* **1980,** *38,* 121.
10.   Poliakoff, M.; Turner, J. J. *J. Chem Soc., Faraday Trans. 2* **1974,** *70,* 93.
11.   King, R. B. *J. Inorg. Nucl. Chem.* **1969,** *5,* 906.
12.   Zeigler, T.; Tschinke, V.; Ursenbach, C. *J. Am. Chem. Soc.* **1987,** *109,* 4825.

13. Brower, K. R.; Chen, T.-S. *Inorg. Chem.* **1973**, *12*, 2198.
14. Wojcicki, A.; Basolo, F. *J. Am. Chem. Soc.* **1961**, *83*, 527.
15. Johnson, B. F. G.; Lewis, J.; Meiser, J. R.; Robinson, B. H.; Robinson, P. W.; Wojcicki, A. *J. Chem. Soc. A* **1968**, 522; Kaesz, H. D.; Bau, R.; Hendrikson, D.; Smith, M. *J. Am. Chem. Soc.* **1967**, *89*, 2844; Robinson, P. W.; Cohen, M. A.; Wojcicki, A. *Inorg. Chem.* **1971**, *10*, 2081; Berry, A.; Brown, T. L. *Inorg. Chem.* **1972**, *11*, 1165.
16. Atwood, J. D.; Brown, T. L. *J. Am. Chem. Soc.* **1975**, *97*, 3380.
17. Chen, L.; Poë, A. *J. Inorg. Chem.* **1989**, *28*, 3641.
18. Schuster-Woldan, H. G.; Basolo, F. *J. Am. Chem. Soc.* **1966**, *88*, 1657.
19. Cramer, R.; Seiwell, L. P. *J. Organomet. Chem.* **1975**, *92*, 245; Bleeke, J. R.; Peng, W. J. *Organometallics* **1986**, *5*, 635.
20. Ji, L.-N.; Rerek, M. E.; Basolo, F. *Organometallics* **1984**, *3*, 740.
21. Kowaleski, R. M.; Basolo, F.; Trogler, W. C.; Ernst, R. D. *J. Am. Chem. Soc.* **1986**, *108*, 6046.
22. Morris, D. E.; Basolo, F. *J. Am. Chem. Soc.* **1968**, *90*, 2531.
23. Darensbourg, D. J.; Darensbourg, M. Y.; Walker, N. *Inorg. Chem.* **1981**, *20*, 1918; Cotton, F. A.; Darensbourg, D. J.; Kolthammer, B. W. S.; Kudaroski, R. *Inorg. Chem.* **1982**, *21*, 1656.
24. Rossi, A. G.; Hoffmann, R. *Inorg. Chem.* **1975**, *14*, 365.
25. Poliakoff, M. *Inorg. Chem.* **1976**, *15*, 2892.
26. Cotton, F. A.; Darensbourg, D. J.; Klein, S.; Kolthammer, B. W. S. *Inorg. Chem.* **1982**, *21*, 294.
27. Ewen, J. E.; Darensbourg, D. J. *J. Am. Chem. Soc.* **1975**, *97*, 6874.
28. Graham, J. R.; Angelici, R. J. *Inorg. Chem.* **1967**, *6*, 992.
29. Memering, M. N.; Dobson, G. R. *Inorg. Chem.* **1973**, *12*, 2490.
30. Halverson, D. E.; Reisner, G. M.; Dobson, G. R.; Bernal, I.; Mulcahy, T. L. *Inorg. Chem.* **1982**, *21*, 4285.
31. Macholdt, H.-T.; van Eldik, R.; Dobson, G. R. *Inorg. Chem.* **1986**, *25*, 1914.
32. Kubas, G. J. *Acc. Chem. Res.* **1988**, *21*, 120.
33. Gonzalez, A. A.; Zhang, K.; Hoff, C. D. *Inorg. Chem.* **1989**, *28*, 4285.
34. Wawersik, H.; Basolo, F. *Inorg. Chim. Acta* **1969**, *3*, 113.
35. Fawcett, J. P.; Poë, A.; Sharma, K. R. *J. Am. Chem. Soc.* **1976**, *98*, 1401; Fawcett, J. P.; Poë, A. *J. Chem. Soc., Dalton Trans.* **1977**, 1302.
36. Wegman, R. W.; Olsen, R. J.; Gard, D. R.; Faulkner, L. R.; Brown, T. L. *J. Am. Chem. Soc.* **1981**, *103*, 6089.
37. Sonnenberger, D.; Atwood, J. D. *J. Am. Chem. Soc.* **1980**, *102*, 3484.
38. Coville, N. J.; Stolzenberg, A. M.; Meutterties, E. L. *J. Am. Chem. Soc.* **1983**, *105*, 2499.
39. Poë, A. *Inorg. Chem.* **1981**, *20*, 4029, 4032.
40. Atwood, J. D. *Inorg. Chem.* **1981**, *20*, 4031.
41. Absi-Halabi, M.; Brown, T. L. *J. Am. Chem. Soc.* **1977**, *99*, 2983.
42. Byers, B. H.; Brown, T. L. *J. Am. Chem. Soc.* **1977**, *99*, 2527.
43. Byers, B. H.; Brown, T. L. *J. Organomet. Chem.* **1977**, *127*, 181.
44. Sweany, R. L.; Halpern, J. *J. Am. Chem. Soc.* **1977**, *99*, 8335.
45. Bullock, R. M.; Samsel, E. G. *J. Am. Chem. Soc.* **1991**, *112*, 6886.
46. Herrinton, T. R.; Brown, T. L. *J. Am. Chem. Soc.* **1985**, *107*, 5700.

47. McCullen, S. B.; Walker, H. W.; Brown, T. L. *J. Am. Chem. Soc.* **1982**, *104*, 4007.
48. Fox, A.; Malito, J.; Poë, A. *J. Chem. Soc., Chem. Commun.* **1981**, 1052.
49. Therien, M. J.; Ni, C.-L.; Anson, F. C.; Osteryoung, J. G.; Trogler, W. C. *J. Am. Chem. Soc.* **1986**, *108*, 4037.
50. Therien, M. J.; Trogler, W. C. *J. Am. Chem. Soc.* **1988**, *110*, 4942.
51. Noack, K.; Calderazzo, F. *J. Organomet. Chem.* **1967**, *10*, 101.
52. Flood, T. C.; Jensen, J. E.; Statler, J. A. *J. Am. Chem. Soc.* **1981**, *103*, 4410.
53. Wright, S. C.; Baird, M. C. *J. Am. Chem. Soc.* **1985**, *107*, 6899.
54. Flood, T. C.; Campbell, K. D. *J. Am. Chem. Soc.* **1984**, *106*, 2853.
55. Brunner, H.; Hammer, B.; Bernal, I.; Draux, M. *Organometallics* **1983**, *2*, 1595.
56. Mawby, R. J.; Basolo, F.; Pearson, R. G. *J. Am. Chem. Soc.* **1964**, *86*, 3994.
57. Mawby, R. J.; Basolo, F.; Pearson, R. G. *J. Am. Chem. Soc.* **1964**, *86*, 5043.
58. Butler, I. S.; Basolo, F.; Pearson, R. G. *Inorg. Chem.* **1967**, *6*, 2074.
59. Wax, M. J.; Bergman, R. G. *J. Am. Chem. Soc.* **1981**, *103*, 7028.
60. Collman, J. P.; Finke, R. G.; Cawse, J. N.; Brauman, J. I. *J. Am. Chem. Soc.* **1978**, *100*, 4766.
61. Wojcicki, A. *Adv. Organomet. Chem.* **1974**, *12*, 31.
62. Whitesides, G. M.; Boschetto, D. J. *J. Am. Chem. Soc.* **1971**, *93*, 1529.
63. Darensbourg, D. J.; Bauch, C. G.; Reibenspies, J. H.; Rheingold, A. L. *Inorg. Chem.* **1988**, *27*, 4203.
64. Darensbourg, D. J.; Hanckel, R. K.; Bauch, C. G.; Pala, M.; Simmons, D.; White, J. N. *J. Am. Chem. Soc.* **1985**, *107*, 7463; Darensbourg, D. J.; Grötsch, G. *J. Am. Chem. Soc.* **1985**, *107*, 7473.
65. Burk, M. J.; McGrath, M. P.; Wheeler, R.; Crabtree, R. H. *J. Am. Chem. Soc.* **1988**, *110*, 5034.
66. Kubas, G. J.; Unkefer, C. J.; Swanson, B. I.; Fukushima, E. *J. Am. Chem. Soc.* **1986**, *108*, 7000, and references therein.
67. Hay, P. J. *J. Am. Chem. Soc.* **1987**, *109*, 705.
68. Saillard, J.-Y.; Hoffmann, R. *J. Am. Chem. Soc.* **1984**, *106*, 2006.
69. Choi, H. W.; Muetterties E. L. *J. Am. Chem. Soc.* **1982**, *104*, 153.
70. Van-Catledge, F. A.; Ittel, S. D.; Jesson, J. P. *Organometallics* **1985**, *4*, 18.
71. Halpern, J.; Wong, C. S. *J. Chem. Soc., Chem. Commun.* **1973**, 629.
72. Shaw, B. L.; Stainbank, R. E. *J. Chem. Soc., Dalton Trans.* **1972**, 223; Miller, E. M.; Shaw, B. L. *J. Chem. Soc., Dalton Trans.* **1974**, 480.
73. Labinger, J. A.; Osborn, J. A. *Inorg. Chem.* **1980**, *19*, 3230; Labinger, J. A.; Osborn. J. A.; Coville, N. J. *Inorg. Chem.* **1980**, *19*, 3236.
74. Williams, G. M.; Schwartz, J. *J. Am. Chem. Soc.* **1982**, *104*, 1122.
75. Basil, J. D.; Murray, H. H.; Fackler, J. P., Jr.; Tocher, J.; Mazany, A. M.; Trzcinska-Bancroft, B.; Knachel, H.; Dudis, D.; Delord, T. J.; Marler, D. O. *J. Am. Chem. Soc.* **1985**, *107*, 6908.
76. Kramer, A. V.; Labinger, J. A.; Bradley, J. S.; Osborn, J. A. *J. Am. Chem. Soc.* **1974**, *96*, 7145; Kramer, A. V.; Osborn, J. A. *J. Am. Chem. Soc.* **1974**, *96*, 7832.

77. Stoutland, P. O.; Bergman, R. G. *J. Am. Chem. Soc.* **1988,** *110,* 5732.
78. Baker, M. V.; Field, L. D. *J. Am. Chem. Soc.* **1986,** *108,* 7436.
79. Ghosh, C. K.; Hoyano, J. K.; Krentz, R.; Graham, W. A. G. *J. Am. Chem. Soc.* **1989,** *111,* 5480.
80. Becker, P. N.; Bergman, R. G. *J. Am. Chem. Soc.* **1983,** *105,* 2985.
81. Osborn, J. A.; Jardine, F. H.; Young, J. F.; Wilkinson, G. *J. Chem. Soc. A* **1966,** 1711; Jardine, F. H.; Osborn, J. A.; Wilkinson, G. *J. Chem. Soc. A* **1967,** 1574; Montelatici, S.; van der Ent, A.; Osborn, J. A.; Wilkinson, G. *J. Chem. Soc. A* **1968,** 1054.
82. Arai, H.; Halpern, J. *J. Chem. Soc., Chem. Commun.* **1971,** 1571.
83. Meakin, P.; Jesson, J. P.; Tolman, C. A. *J. Am. Chem. Soc.* **1972,** *94,* 3240.
84. Halpern, J.; Wong, S. W. *J. Chem. Soc., Chem. Commun.* **1973,** 629.
85. Tolman, C. A.; Meakin, P. Z.; Lindner, D. L.; Jesson, J. P. *J. Am. Chem. Soc.* **1974,** *96,* 2762.
86. Halpern, J.; Okamoto, T.; Zakhariev, A. *J. Mol. Catal.* **1977,** *2,* 65.
87. Halpern, J.; Riley, D. P.; Chan, A. S. C.; Pluth, J. J. *J. Am. Chem. Soc.* **1977,** *99,* 8055.
88. Chan, A. S. C.; Pluth, J. J.; Halpern, J. *J. Am. Chem. Soc.* **1980,** *102,* 5952.
89. Landis, C. R.; Halpern, J. *J. Am. Chem. Soc.* **1987,** *109,* 1746.
90. Brunner, H. *Top. Stereochem.* **1988,** *18,* 129; Noyori, R. *Chem. Soc. Rev.* **1989,** *18,* 187.
91. Takahashi, H.; Sakuraba, S.; Takeda, H.; Achiwa, K. *J. Am. Chem. Soc.* **1990,** *112,* 5876.
92. Chan, Y. N. C.; Osborn, J. A. *J. Am. Chem. Soc.* **1990,** *112,* 9400.
93. Garland, M.; Blaser, H.-U. *J. Am. Chem. Soc.* **1990,** *112,* 7048.
94. Janowicz, A. H.; Bergman, R. G. *J. Am. Chem. Soc.* **1982,** *104,* 352.
95. Hoyano, J. K.; Graham, W. A. G. *J. Am. Chem. Soc.* **1982,** *104,* 3723.
96. Ghosh, C. K.; Graham, W. A. G. *J. Am. Chem. Soc.* **1987,** *109,* 4726.
97. Ghosh, C. K.; Graham, W. A. G. *J. Am. Chem. Soc.* **1989,** *111,* 375.
98. Brookhart, M.; Green, M. L. H.; Wong, L.-L. *Prog. Inorg. Chem.* **1988,** *36,* 1.
99. Crabtree, R. H. *Chem. Rev.* **1985,** *85,* 245.
100. Rothwell, I. P. *Acc. Chem. Res.* **1988,** *21,* 153.
101. Thompson, M. E.; Bercaw, J. E. *Pure Appl. Chem.* **1984,** *56,* 1.
102. Nolan, S. P.; Hoff, C. D.; Stoutland, P. D.; Newman, L. J.; Buchanan, J. M.; Bergman, R. G.; Yang, G. K.; Peters, K. S. *J. Am. Chem. Soc.* **1987,** *109,* 3143.
103. Marx, D. E.; Lees, A. J. *Inorg. Chem.* **1988,** *27,* 1121.
104. Drolet, D. P.; Lees, A. J. *J. Am. Chem. Soc.* **1990,** *112,* 5878.
105. Weiller, B. H.; Wasserman, E. P.; Bergman, R. G.; Moore, C. B.; Pimentel, G. C. *J. Am. Chem. Soc.* **1989,** *111,* 8288.
106. Belt, S. T.; Grevels, F.-W.; Klotzbücher, W. E.; McCamley, A.; Perutz, R. N. *J. Am. Chem. Soc.* **1989,** *111,* 8373.
107. Bullock, R. M.; Headford, C. E. L.; Hennessy, K. M.; Kegley, S. E.; Norton, J. R. *J. Am. Chem. Soc.* **1989,** *111,* 3897.
108. Buchanan, J. M.; Stryker, J. M.; Bergman, R. G. *J. Am. Chem. Soc.* **1986,** *108,* 1537.

109. Pruett, R. L. *Adv. Organomet. Chem.* **1979**, *17*, 1.
110. Heck, R. F.; Breslow, D. S. *J. Am. Chem. Soc.* **1961**, *83*, 4023.
111. Yagupsky, G.; Brown, C. K.; Wilkinson, G. *J. Chem. Soc. A* **1970**, 1392; Brown, C. K.; Wilkinson, G. *J. Chem. Soc. A* **1970**, 2753.
112. Kang, H.; Mauldin, C. H.; Cole, T.; Slegeir, W.; Cann, K.; Pettit, R. *J. Am. Chem. Soc.* **1977**, *99*, 8323.
113. Thomas, M. G.; Beier, B. F.; Muetterties, E. L. *J. Am. Chem. Soc.* **1976**, *98*, 1296; Demitras, G. C.; Muetterties, E. L. *J. Am. Chem. Soc.* **1977**, *99*, 2796.
114. Laine, R. M.; Rinker, R. G.; Ford, P. C. *J. Am. Chem. Soc.* **1977**, *99*, 252.
115. Masters, C. *Adv. Organomet. Chem.* **1979**, *17*, 61.
116. Cramer, R. *Acc. Chem. Res.* **1968**, *1*, 186.
117. Tolman, C. A. *Chem. Rev.* **1977**, *77*, 313.

6

Oxidation–Reduction Reactions

For most purposes, inorganic reactions can be classified as either substitution reactions or oxidation–reduction reactions. The latter involve the transfer of at least one electron from the reducing agent to the oxidizing agent. Such reactions are widely used in analytical procedures and are important in many biological processes. Oxidation–reduction reactions have been classified in two general ways; the first historically is by stoichiometry and the second is by mechanism.

6.1 CLASSIFICATION OF REACTIONS

6.1.a Stoichiometric Classification

The stoichiometric classification is a classical method that only requires a knowledge of the reaction stoichiometry but has limited kinetic applicability.

6.1.a.i Complementary Reactions

The change in oxidation state of the reducing agent is the same as the change in oxidation state of the oxidizing agent. Some examples are

$$Cr^{2+} + Ag^+ \longrightarrow Ag^0 + Cr^{3+}$$

$$Zn^0 + Cu^{2+} \longrightarrow Cu^0 + Zn^{2+}$$

(6.1)

6.1.a.ii Noncomplementary Reactions

The oxidizing agent and the reducing agent undergo different net changes in oxidation state. Some examples are

$$2\,Cr^{2+} + Tl^{3+} \longrightarrow 2\,Cr^{3+} + Tl^+$$

$$Zn^0 + 2\,Fe^{3+} \longrightarrow Zn^{2+} + 2\,Fe^{2+}$$

(6.2)

This classification has no direct mechanistic implications. However, it is a qualitative observation that complementary reactions are faster than noncomplementary reactions. This "rule" can be useful in designing analytical and preparative procedures, but it is by no means universal. The mechanistic rationale for this qualitative kinetic observation is based on the assumption that the reactions occur in bimolecular steps. Then noncomplementary reactions must normally proceed through an unstable oxidation state of one of the reactants, as in the following examples:

$$Cr^{2+} + Tl^{3+} \longrightarrow Cr^{3+} + \{Tl^{2+}\}$$

$$Cr^{2+} + \{Tl^{2+}\} \longrightarrow Cr^{3+} + Tl^{+}$$

(6.3)

or

$$Cr^{2+} + Tl^{3+} \longrightarrow \{Cr^{4+}\} + Tl^{+}$$

$$\{Cr^{4+}\} + Cr^{2+} \longrightarrow 2\ Cr^{3+}$$

(6.4)

For the reaction sequence in Eq. (6.3), Tl^{2+} is an unstable oxidation state of thallium, and in Eq. (6.4), Cr^{4+} is an unstable state for chromium.

6.1.b   Mechanistic Classification

The mechanistic classification obviously requires a knowledge of the reaction mechanism. A major theme of this area is the elucidation of the kinetic features that allow one to determine the type of mechanism.

6.1.b.i   Inner-Sphere Electron Transfer

At least one ligand is shared in the first coordination sphere of the oxidant and reductant in the transition state for the electron-transfer process:

$$L_5M^{III}X + M^{II}Y_6 \longrightarrow \{L_5M^{III}\!-\!X\!-\!M^{II}Y_5\} \longrightarrow \text{Products} \qquad (6.5)$$
$$+\ Y$$

The formation of the transition state involves a substitution reaction on one of the metal centers in order to form the intermediate or transition state with the bridging ligand —X— in Eq. (6.5). Then electron transfer occurs, possibly via the bridging ligand.

This mechanism was first demonstrated by Taube and co-workers[1] in a system that exploited the ideas of substitution lability and inertness that Taube was developing at the same time. Complexes of cobalt(III) (low-spin d^6) are inert, of chromium(II) (high-spin d^4) are labile, of cobalt(II) (high-spin d^7) are labile, and of chromium(III) (d^3) are inert. This combination allowed the reaction products to establish definitively an inner-sphere mechanism for the reaction in Scheme 6.1. The metal ions retain their original oxidation states in the precursor complex. The relative labilities of Co(II) and Cr(III) cause the successor complex to decompose to the observed stable Cr(III) product in which the Cl$^-$ ligand originally on cobalt(III) has been transferred to chromium(III).

Scheme 6.1

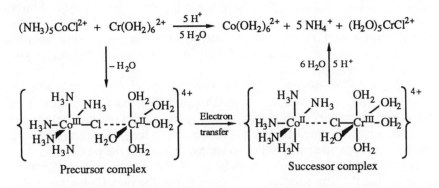

$$(NH_3)_5CoCl^{2+} + Cr(OH_2)_6^{2+} \xrightarrow[5\,H_2O]{5\,H^+} Co(OH_2)_6^{2+} + 5\,NH_4^+ + (H_2O)_5CrCl^{2+}$$

Precursor complex   Successor complex

The example in Scheme 6.1 was crucial to establishing that inner-sphere or bridged electron transfer can occur, but only rarely does one have the proper combination of substitution lability patterns so that the products uniquely define the mechanism. More commonly, this mechanism is inferred if the electron transfer is unusually fast and sensitive to the chemical nature of the bridging group. A wide range of bridging ligands is known, such as the halide ions, pseudohalide ions, hydroxide ion, and carboxylate ions. Ammonia cannot act as a bridging ligand because it does not have a second unshared pair of electrons, which is necessary for simultaneous bonding to the oxidizing and reducing centers. The situation with H_2O is unclear in that it has a second electron pair but does not produce much kinetic acceleration. The effect of the bridging group is discussed later in the chapter.

6.1.b.ii Outer-Sphere Electron Transfer

The coordination spheres of the oxidant and reductant remain intact during the electron transfer, as shown in the following reaction:

$$M^{III}L_6 + M^{II}Y_6 \longrightarrow (L_5M^{III}L)(YM^{II}Y_5) \longrightarrow M^{II}L_6 + M^{III}Y_6 \quad (6.6)$$

The reactants are considered as hard charged spheres and an electrostatic approach can be used to anticipate the rates of these reactions. To date, this is one of the few areas in which theory has provided useful guidelines for mechanistic studies. A mechanistic decision between the inner- and outer-sphere possibilities is often based on whether or not the reaction rate corresponds reasonably to the predictions of the outer-sphere theory.

6.2 OUTER-SPHERE ELECTRON-TRANSFER THEORY

The outer-sphere theory has been developed using an electrostatic approach to calculate the energy necessary to bring reactants together, to reorganize the solvent around the transition state, and to prepare the metal centers for electron transfer.

In North America the theory is associated with the name of Marcus and referred to as Marcus's theory.[2] However, Hush[3] in Australia and Levitch[4] and Dogonadze[5] in the U.S.S.R. have made original contributions. Marcus started from the transition-state theory for ionic reactions, Hush from solid-state electron-transfer theory, and Levitch from a consideration of reactions at electrodes. All arrived at essentially the same result although using different terminologies.

The *Franck–Condon principle* is fundamental to the theory. This principle states that electron movement is much faster than nuclear motion; thus, internuclear distances do not change during the instant of electron transfer. Therefore, it is assumed that on approaching the transition state, the bond lengths of the reactants will adjust to approach those of the products.

The electron transfer is assumed to be an *adiabatic process in the Ehrenfest sense* so that the transmission coefficient (ρ) in the transition-state theory expression, Eq. (6.7), is equal to one.

$$k = \rho \frac{kT}{h} \exp\left(-\frac{\Delta H^*}{RT} + \frac{\Delta S^*}{R} \right) \qquad (6.7)$$

This implies that there is enough interaction between reactants in the transition state to make the probability of electron transfer equal to one. There are numerous occasions when this condition is thought not to be entirely true, and the effect of "nonadiabaticity" on the reaction rate is sometimes used as a rationale for differences between observed and predicted rate constants.

6.2.a Marcus Cross Relationship from Thermodynamics

One of the most important results to evolve from the theoretical treatment of Marcus is now referred to as the Marcus cross relationship. This important relationship was developed later by Ratner and Levine[6] from a thermodynamic perspective, and this formulation provides a simple basis for understanding some of the concepts and assumptions in the more microscopic molecular theory of Marcus.

The electron-transfer reaction between a reductant A^- and oxidant B is given by the following net reaction:

$$A^- + B \longrightarrow A + B^- \qquad (6.8)$$

It is assumed that the reactants come together to form a precursor complex

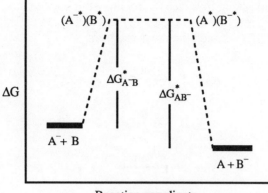

Figure 6.1. Reaction coordinate diagram for an outer-sphere electron-transfer reaction.

$(A^{-*})(B^*)$; then electron transfer occurs to give the successor complex $(A^*)(B^{-*})$, which decomposes to products, as shown in the reaction coordinate diagram in Figure 6.1.

Ratner and Levine assumed that in the precursor and successor complexes, one can define thermodynamic properties for the individual partners (A^{-*}), (B^*), (A^*), and (B^{-*}). This amounts to assuming that there is no significant bonding between the partners in the precursor and successor complexes.

The free energies of activation for the forward and reverse reactions, respectively, are

$$\Delta G^*_{A^-B} = G^o(A^{-*}) + G^o(B^*) - G^o(A^-) - G^o(B) \tag{6.9}$$

$$\Delta G^*_{AB^-} = G^o(A^*) + G^o(B^{-*}) - G^o(A) - G^o(B^-) \tag{6.10}$$

The net free energy change for the reaction is

$$\Delta G^o_{A^-B} = G^o(A) + G^o(B^-) - G^o(A^-) - G^o(B)$$
$$= \Delta G^*_{A^-B} - \Delta G^*_{AB^-} \tag{6.11}$$

If Eqs. (6.9) and (6.10) are used to substitute for the two terms on the right in Eq. (6.11), it is found that

$$G^o(A^{-*}) + G^o(B^*) = G^o(A^*) + G^o(B^{-*}) \tag{6.12}$$

Thus, the free energies of the precursor and successor complexes are equal.

The cross relationship involves the relationship between the rate constants for the following reactions, where (6.13) and (6.14) are called self-exchange reactions and (6.15) is called the cross reaction:

$$A^- + A \xrightarrow{\ k_{AA}\ } A + A^- \tag{6.13}$$

$$B + B^- \xrightarrow{\ k_{BB}\ } B^- + B \tag{6.14}$$

$$A^- + B \xrightarrow{\ k_{AB}\ } A + B^- \tag{6.15}$$

It is assumed that the free energies of the individual partners in the self-exchange reactions are the same as in the cross reaction; therefore

$$\Delta G_{AA}^* = G^o(A^{-*}) + G^o(A^*) - G^o(A^-) - G^o(A) \tag{6.16}$$

$$\Delta G_{BB}^* = G^o(B^{-*}) + G^o(B^*) - G^o(B^-) - G^o(B) \tag{6.17}$$

$$\Delta G_{AB}^* = G^o(A^{-*}) + G^o(B^*) - G^o(A^-) - G^o(B) \tag{6.18}$$

If Eq. (6.18) is multiplied by 2 and substitution is made from Eq. (6.12), one obtains

$$\begin{aligned} 2\,\Delta G_{AB}^* = {}& G^o(A^{-*}) + G^o(B^*) + G^o(A^*) + G^o(B^{-*}) \\ & - 2\,G^o(A^-) - 2\,G^o(B) \end{aligned} \tag{6.19}$$

which, after rearrangement and substitution from Eqs. (6.16) and (6.17), gives

$$\Delta G_{AB}^* = 1/2 \left(\Delta G_{AA}^* + \Delta G_{BB}^* + \Delta G_{A-B}^o \right) \tag{6.20}$$

This is the Marcus cross relationship in terms of free energies. The thermodynamic development of the cross relationship depends on the *assumptions* that:
1. The activation process for each reactant is independent of the other reactant.
2. The activated species are the same for the self-exchange and the cross reaction.

Clearly, these assumptions will not be valid for an inner-sphere mechanism. They will also be invalid if there are special attractive or repulsive forces between A and B that are not present between A and A^- or B and B^-.

From the transition-state theory, the free energy of activation and the rate constant are related by

$$k_{ij} = Z_{ij} \exp\left(-\frac{\Delta G_{ij}^*}{RT}\right) \tag{6.21}$$

where Z_{ij} is the collision frequency. If this expression and the thermodynamic relationship $\Delta G_{A-B}^o = -RT \ln K_{AB}$ are substituted into Eq. (6.20), then one obtains

$$k_{AB} = \left(k_{AA} k_{BB} K_{AB} \frac{Z_{AB}^2}{Z_{AA} Z_{BB}}\right)^{1/2}$$

$$= (k_{AA} k_{BB} K_{AB} F_{AB})^{1/2} \tag{6.22}$$

This is the cross relationship in terms of rate constants. There is often reason to believe (or need to assume) that $F_{AB} \approx 1$, and then Eq. (6.22) reduces to what is often called the simplified Marcus cross relationship. This is particularly useful because a knowledge of any three of the values, k_{AB}, k_{AA}, k_{BB}, or K_{AB} allows one to predict the fourth.

The assumption that $F_{AB} \approx 1$, means that $Z_{AB}^2 \approx Z_{AA} Z_{BB}$. For ionic reactants, this would seem quite reasonable if the reaction has charge symmetry (e.g., $A^{2+} + B^{3+} \rightarrow A^{3+} + B^{2+}$) since the charges of the species in the self-exchange reactions are the same as those in the net reaction. The assumption is somewhat less valid if the species are of the same charge type but the reaction lacks charge symmetry (e.g., $A^{2+} + B^{3+} \rightarrow A^{1+} + B^{4+}$). In this example, Z_{AA} for $A^{2+} + A^{1+}$ will be larger than Z_{BB} for $B^{3+} + B^{4+}$ whereas Z_{AB} for $A^{2+} + B^{3+}$ may be intermediate between the two, with the net result that $Z_{AB}^2 \approx Z_{AA} Z_{BB}$ may still be true within a factor of 10. However, if the reactants are of opposite charge type (e.g., $A^{2-} + B^{3+} \rightarrow A^{3-} + B^{2+}$), then the attractive force between the ions of opposite charge will make Z_{AB} larger than either Z_{AA} or Z_{BB} and F_{AB} will be much larger than one.

The preceding development and discussion serve as a preamble to the more detailed development that follows. The detailed theory takes into account the charge asymmetry problems by calculating work terms for bringing reactants together and separating the products. When the theoretical predictions still seem inadequate, nonadiabaticity is often invoked; this is equivalent to attributing the deviations to the Z values.

6.2.b   Marcus Theory Details

The details of the Marcus theory have been described in several reviews[7–9] and in books by Reynolds and Lumry[10] and Cannon.[11] The following discussion will simply outline the features of the theory and give the physical factors that are predicted to be important in determining the rates of outer-sphere electron-transfer reactions.

The reactants are considered to be two hard spheres of charge z_1 and z_2 and radii a_1 and a_2. Work will be required to bring the reactants together to a separation $r = a_1 + a_2$, which is considered to be the reactant separation in the transition state. From simple electrostatics, this coulombic work contribution to the free energy of activation is given by

$$\Delta G_{coul}^* = \left(\frac{N z_1 z_2 e^2}{4 \pi \varepsilon_0} \right) \left(\frac{1}{\varepsilon_s r} \right) \tag{6.23}$$

where e is the charge on the electron and ε_s is the bulk dielectric constant of the solvent. In more recent treatments, the formation of the precursor complex has been considered as a diffusion-controlled equilibrium reaction, so that the ratio of the calculated forward and reverse rate constants gives Eq. (6.24), the equilibrium constant for outer-sphere precursor complex formation,

$$K_{os} = \frac{4 \pi N r^3}{3000} \exp\left(\frac{U}{1 + B r \sqrt{\mu}} \right) \tag{6.24}$$

where

$$U = \frac{z_1 z_2 e^2}{4 \pi \varepsilon_0 \varepsilon_s r k T}$$

and the terms in U have been defined previously in Eq. (1.71). The B is the Debye–Hückel factor discussed in Eq. (1.70) and μ is the ionic strength. The latter terms obviously are introduced in an attempt to correct for ionic strength variations. Alternative approaches to the ionic strength effect have been described in Chapter 1 and by Tembe et al.[12]

When the reactants come together, they are considered to form a spherical transition state of diameter r. The solvent molecules will reorganize around the transition state, and this solvent or outer-sphere reorganization contribution is given by

$$\Delta G_{solv}^* = (\Delta q)^2 \left(\frac{N e^2}{16 \pi \varepsilon_0} \right) \left(\frac{1}{2 a_1} + \frac{1}{2 a_2} - \frac{1}{r} \right) \left(\frac{1}{n_s^2} - \frac{1}{\varepsilon_s} \right) \tag{6.25}$$

where Δq is the number of electrons transferred (one for the majority of reactions) and n_s is the refractive index of the solvent (n_s^2 is the high-frequency dielectric constant of the solvent).

In order to satisfy the Franck–Condon principle, the ligands around the metal ions will adjust their bond lengths in the transition state toward the lengths they will have in the products. This factor is called the inner-sphere reorganization energy and is given by

$$\Delta G_{in}^{*} = N\left(\frac{n f_1}{2}\left(d_1^o - d_1^*\right)^2 + \frac{n f_2}{2}\left(d_2^o - d_2^*\right)^2\right) \quad (6.26)$$

where n is the number of ligands, f_1 and f_2 are the force constants for the symmetrical breathing vibrational mode, which is assumed to generate the appropriate bond lengthening or shortening, and d_i^o and d_i^* are the ground-state and transition-state metal–ligand bond lengths, respectively. The ΔG_{in}^{*} is the most difficult term to calculate because d_i^* is unknown and the force constants require an assignment of the vibrational spectrum in the difficult region of ~200 to 800 cm^{-1}. One approach is to let $d^* = d_1^* = d_2^*$ and then solve for d^* when ΔG_{in}^{*} is a minimum [i.e., when $d(\Delta G_{in}^{*})/d(d^*) = 0$]. Then

$$d^* = \frac{f_1 d_1^o + f_2 d_2^o}{f_1 + f_2} \quad (6.27)$$

and

$$\Delta G_{in}^{*} = N\left(\frac{n f_1 f_2}{2(f_1 + f_2)}\right)\left(d_2^o - d_1^o\right)^2 \quad (6.28)$$

To determine the force constants f_i, the appropriate potential energy function (U) should be used to calculate $f = d^2(U)/d^2(d)$. A simple harmonic diatomic oscillator is often assumed, so that $f_i = (2\pi v_i c)^2 m_r$, where v_i is the vibrational frequency in cm^{-1}, c is the speed of light in cm s^{-1}, and m_r is the reduced mass of the ligand in kilograms.

Nuclear tunneling has also been included as a pre-exponential factor (Γ_n) in the rate constant expression. The nuclear vibrational frequency of the reactants (v_n) is assumed to be given by

$$v_n^2 = \frac{v_{solv}^2 \Delta G_{solv}^{*} + v_{in}^2 \Delta G_{in}^{*}}{\Delta G_{solv}^{*} + \Delta G_{in}^{*}} \approx \frac{v_{in}^2 \Delta G_{in}^{*}}{\Delta G_{solv}^{*} + \Delta G_{in}^{*}} \quad (6.29)$$

where the approximation uses the fact that the frequency for the outer-sphere solvent molecules (v_{solv}), ~30 cm^{-1}, is more than 10 times smaller than the inner-sphere frequency (v_{in}), ~400 cm^{-1}, and $\Delta G_{in}^{*} \approx \Delta G_{solv}^{*}$. The value of v_{in} is taken as the average of the breathing-mode frequencies of the two reactants as calculated from

$$v_{in}^2 = \frac{2 v_1^2 v_2^2}{v_1^2 + v_2^2} \tag{6.30}$$

With v_n in cm^{-1} and $c = 3 \times 10^{10}$ cm s^{-1}, $\Gamma_n = 3 \times 10^{10} v_n$.

The final feature to be added is the electronic transmission coefficient (κ_{el}), which is equal to one if the electron-transfer step is adiabatic and is smaller otherwise. Then the calculated outer-sphere rate constant is given by Eq. (6.31):

$$k_{calc} = \kappa_{el} \Gamma_n K_{os} \exp\left(\frac{-\left(\Delta G_{solv}^* + \Delta G_{in}^*\right)}{RT}\right) \tag{6.31}$$

Clearly, the next step is to compare calculated rate constants from Eq. (6.31) to some experimental results. Because of their overall symmetry, it is easier to do the calculation for electron-exchange reactions such as (6.32) where M and L are the same and only the oxidation state is different in the two reactants:

$$M(L)_n^{z+} + {}^*M(L)_n^{(z-1)+} \longrightarrow M(L)_n^{(z-1)+} + {}^*M(L)_n^{z+} \tag{6.32}$$

Results for some such systems are given in Table 6.1. The results in Table 6.1 show that the main feature controlling the relative rates for different metal ions is the differences in ΔG_{in}^* which are in turn controlled by the differences in bond lengths of the two oxidation states. The agreement of the calculated and observed values is quite good for the Fe and Co systems but poor for $Ru(OH_2)_6^{2/3+}$. Clearly, some approximations have gone into these calculations in addition to those involved in the theory. The most critical factor is the bond length or size difference between the two oxidation states because ΔG_{in}^* depends on the square of this difference. For example, the observed and calculated values for $Ru(OH_2)_6^{2/3+}$ agree if the size difference is 0.125 Å instead of 0.09 Å, and a change from 0.04 Å to 0.095 Å gives agreement for $Ru(NH_3)_6^{2/3+}$. The force constants used in the calculations might also be questioned because of the simplicity of the model, although the errors will be somewhat self-compensating, as can be seen from the form of Eq. (6.28). Bernhard and Ludi[13] have done a normal coordinate analysis on the $Ru(OH_2)_6^{2/3+}$ system and obtained values of 191 and 298 N m^{-1} for the 2+ and 3+ oxidation states, respectively, and the simple model based on a harmonic oscillator for the symmetric breathing mode gives 190 and 300 N m^{-1}, respectively. This factor does not seem to be a source of concern. The term most commonly blamed for lack of agreement between k_{calc} and k_{obsd} is the electronic transmission coefficient (κ_{el}) because calculated values are usually higher than observed, so that nonadiabaticity is invoked with κ_{el} in the range of 0.1 to 10^{-3} to bring the values into agreement.

Table 6.1. Comparison of Relevant Parameters and Calculated and Observed Self-Exchange Rate Constants for $M(L)_n^{2/3+}$ Systems

Parameter	$Fe(OH_2)_6^{2/3+}$	$Ru(OH_2)_6^{2/3+}$	$Ru(NH_3)_6^{2/3+}$	$Co(NH_3)_6^{2/3+}$
a_1, a_2 (Å)	3.33, 3.19	3.32, 3.23	3.37, 3.33	3.41, 3.19
v_1, v_2 (cm^{-1})	390, 490	424, 532	442, 500	357, 394
v_n (cm^{-1})	319	286	151	347
$10^{-12} \times \Gamma_n$ (s^{-1})	9.6	8.6	4.5	10.4
μ (M)	0.55	5.0	0.10	2.5
K_{os} (M^{-1})	0.055	0.23	0.017	0.093
ΔG_{solv}^* (kJ mol^{-1})	29.3	29.2	28.5	29.0
ΔG_{in}^* (kJ mol^{-1})	35.0	17.1	3.2	73.4
k_{calc} (M^{-1} s^{-1}, 25°C)	2.8	1.5×10^5	2.2×10^5	1.1×10^{-6}
k_{obsd} (M^{-1} s^{-1}, 25°C)	4.2 [a]	20 [b]	6.7×10^3 [c]	8×10^{-6} [d]

[a] Brunschwig, B. S.; Creutz, C.; Macartney, D. H.; Sham, T.-K.; Sutin, N. *Disc. Faraday Soc.* **1982**, *74*, 113; Jolley, W. H.; Stranks, D. R.; Swaddle, T. W. *Inorg. Chem.* **1990**, *29*, 1948.

[b] Bernhard, P.; Helm, L.; Ludi, A.; Merbach, A. E. *J. Am. Chem. Soc.* **1985**, *107*, 312.

[c] Extrapolated with $\Delta H^* = 5$ kcal mol^{-1} from measurements at 4°C, Smolenaers, P. J.; Beattie, J. K. *Inorg. Chem.* **1986**, *25*, 2259.

[d] At 40°C with trifluoromethanesulfonate counterion, Hammershoi, A.; Geselowitz, D.; Taube, H. *Inorg. Chem.* **1984**, *23*, 979.

The theory also predicts a variation in rate constant with the solvent properties due to the $(n_s^{-2} - \varepsilon_s^{-1})$ dependence of ΔG_{solv}^*. Chan and Wahl[14] found that this prediction was reasonably followed for the tris(hexafluoroacetylacetonato)ruthenium (II)/(III) self-exchange. Recent studies[15] on the $Fe(C_5H_5)_2^{0/+}$ system also show the expected general trend, although the correlation is not perfect. Lay et al.[16] have studied the effect of solvent on the reduction potentials of several cobalt(III) complexes and find a correlation with hydrogen-bonding basicities which indicates the limitations of the conventional continuum model for solvent effects. A similar conclusion might be drawn from Swaddle's detailed analysis[17] of the solvent effects on the volumes of activation for electron-exchange reactions.

The self-exchange rates and reduction potentials for some other metal ion and ligand systems are given in Table 6.2. For the cobalt(II)/(III) complexes in Table 6.2, the k_{AA} values are quite sensitive to the ligand environment and are unusually small when compared to the Ru(II)/(III) analogues. Initially, it was thought that the cobalt systems were slow because of a large inner-sphere reorganization energy caused by the greater bond length differences in the $M(NH_3)_6^{2/3+}$ complexes of cobalt (Co^{2+} 2.11 Å, Co^{3+} 1.94 Å) as

Table 6.2. Reduction Potentials and Self-Exchange Rate Constants (25°C) for Selected Reagents

Redox Couple	$E°$ (V)[a]	k_{AA} (M^{-1} s^{-1})[b]
$V(OH_2)_6^{2/3+}$	−0.255	1×10^{-2}
$Cr(OH_2)_6^{2/3+}$	−0.40	~1×10^{-5}
$Fe(OH_2)_6^{2/3+}$	+0.74	4.0
$Co(OH_2)_6^{2/3+}$	+1.96	~5
$Co(en)_3^{2/3+}$	−0.13	2.0×10^{-5}
$Co(phen)_3^{2/3+}$	+0.42	4×10^1
$Co(sep)^{2/3+}$	−0.26	5.1
$Co(NH_3)_6^{2/3+}$	+0.058	8×10^{-6} [c, d]
$Ru(NH_3)_6^{2/3+}$	+0.051	6.7×10^3 [c]
$Ru(en)_3^{2/3+}$		1.7×10^4 [e]
$Ru(bpy)_3^{2/3+}$	+1.26	2.0×10^9

[a] Creaser, I. I.; Sargeson, A. M.; Zanella, A. W. *Inorg. Chem.* **1983**, *22*, 4022.

[b] See Chou, M.; Creutz, C.; Sutin, N. *J. Am. Chem. Soc.* **1977**, *99*, 5615 for original references, unless otherwise indicated.

[c] See Table 6.1.

[d] At 40°C.

[e] Beattie, J. K.; Smolenaers, P. J. *J. Phys. Chem.* **1986**, *90*, 3684.

compared to ruthenium (Ru^{2+} 2.14 Å, Ru^{3+} 2.10 Å). This results largely because $Co(NH_3)_6^{2+}$ is high-spin. Electronic factors have also been invoked based on the idea that Co must transfer an e_g electron [Co(III) (t_{2g}^6), Co(II) $(t_{2g}^5 e_g^2)$], whereas ruthenium transfers a t_{2g} electron [Ru(III) (t_{2g}^5), Ru(II) (t_{2g}^6)]. A further possibility for the cobalt system has been suggested in which cobalt(II) may need to undergo a spin state change[18] before electron transfer can occur, as in Co(II) $(t_{2g}^5 e_g^2) \rightarrow$ Co(II) $(t_{2g}^6 e_g^1)$. These arguments seem to be qualitatively consistent with the fact that $Co(phen)_3^{2+}$ is close to the low-spin state and has a larger k_{AA} than the hexaammine system.

The situation has been clarified by recent studies of Sargeson and co-workers on macrocyclic cobalt and ruthenium complexes. The self-exchange rate at 25°C[19] for Co(sepulchrate)$^{2/3+}$ is 5.1 M^{-1} s^{-1} in 0.2 M NaCl and 4.8 M^{-1} s^{-1} in 0.2 M LiClO$_4$ although the spin state and bond lengths (Co(II)—N, 2.16 Å; Co(III)—N, 1.96 Å) are essentially the same as in the $Co(NH_3)_6^{2/3+}$ system (sepulchrate is a cage hexadentate N-donor ligand). The much larger rate constant (~10^6 times) has been attributed to strain in the cage ligand because the Co(III) is a bit too small to fit into the cage and the Co(II) is a bit too large.[20] As a result, bond length distortion toward the transition state is particularly favorable. Similar factors may apply to $Co(tacn)_2^{2/3+}$ (an N_3 macrocycle) with $k = 0.18$ M^{-1} s^{-1}, compared to $k = 5 \times 10^4$ M^{-1} s^{-1} for $Ru(tacn)_2^{2/3+}$.[21] For $Co(ttacn)_2^{2/3+}$ (an S-donor chelate) and $Co(azacapten)^{2/3+}$ (an S_3N_3 cage) the bond lengths are almost the same in the two oxidation states because the Co(II) is low-spin, and the rate constants are 1.3×10^4 and 4.5×10^3 M^{-1} s^{-1}, respectively.[22,23] In summary, it appears that bond length differences are a major influence on the self-exchange rates for the cobalt(II)/(III) systems.

6.2.c Applications of the Marcus Cross Relationship

From the detailed theory, Marcus recognized certain simplifications that led to a cross relationship of the same form as that developed by Ratner and Levine. In Marcus's terms, this relationship is given by

$$k_{AB} = (k_{AA} k_{BB} K_{AB} f_{AB})^{1/2} \tag{6.33}$$

where

$$\log f_{AB} = \frac{(\log K_{AB})^2}{4 \log \left(\dfrac{k_{AA} k_{BB}}{Z^2} \right)}$$

and Z is the collision number for the ions in solution ($\approx 10^{11}$ M^{-1} s^{-1}). The f_{AB} factor is analogous to the F_{AB} in the Ratner and Levine development, and $f_{AB} \approx 1$ unless K_{AB} is large. If $f_{AB} \approx 1$, then Eq. (6.33) reduces to Eq. (6.34), called the simplified Marcus equation:

$$k_{AB} = (k_{AA} k_{BB} K_{AB})^{1/2} \tag{6.34}$$

For a one-electron transfer at 25°C, the logarithmic form of this equation is given by

$$\log(k_{AB}) = 0.5 \left[\log(k_{AA}) + \log(k_{BB}) \right] + 0.5 (16.9) \Delta E^\circ \tag{6.35}$$

where ΔE° is the reduction potential in volts. Sutin and co-workers[24] tested this equation using a series of $Fe(phenX)_3^{2/3+}$ complexes, for which k_{BB} is expected to be constant, reacting with aqueous iron(II) and cerium(IV). They found that plots of $\log(k_{AB})$ versus ΔE° have close to the expected slope, but the intercepts are smaller than predicted.

This equation can be applied to calculate one of the self-exchange rate constants (k_{AA} or k_{BB}) or to calculate k_{AB} in order to compare it to an experimental value. These applications have been reviewed by Sutin and co-workers[25] who conclude that k_{AB} can be predicted to within a factor of about 25 if the reaction is outer-sphere. Since the self-exchange rate

Table 6.3. Comparison of Some Observed Rate Constants (M^{-1} s^{-1}, 25°C) with Those Calculated from the Marcus Cross Relationship

Reactants	$\log K_{AB}$	$k_{AB\,obsd}$	$k_{AB\,calc}$
$Ru(NH_3)_6^{2+} + Co(phen)_3^{3+}$	6.25	1.5×10^4	3.5×10^5
$V(OH_2)_6^{2+} + Co(en)_3^{3+}$	2.12	5.8×10^{-4}	3.1×10^{-3}
$V(OH_2)_6^{2+} + Ru(NH_3)_6^{3+}$	5.19	1.3×10^3	2.2×10^3
$V(OH_2)_6^{2+} + Fe(OH_2)_6^{3+}$	16.9	1.8×10^4	1.6×10^6
$Ru(NH_3)_6^{2+} + Fe(OH_2)_6^{3+}$	11.7	3.4×10^5	1.2×10^7
$Fe(OH_2)_6^{2+} + Ru(bpy)_3^{3+}$	8.81	7.2×10^5	3.6×10^8

constant depends on the square of k_{AB}, one could expect to predict k_{AA} or k_{BB} to within 25^2. Some typical results from the cross relationship are given in Table 6.3. The first three examples are typical of the type of agreement that is generally viewed as acceptable. However, the calculations involving the Fe(II)/Fe(III) couple, with $k_{AA} = 4$ M^{-1} s^{-1}, consistently give predicted values that are larger than the experimental values. This effect will be discussed further with regard to the results in Table 6.4.

Bernhard and Sargeson[26] have applied Eq. (6.33) to determine the self-exchange rates of some encapsulated ruthenium, manganese, iron, and nickel complexes. They used five reagents with known self-exchange rates and studied 19 cross reactions to determine five new self-exchange rates by using a least-squares fit of the data to Eq. (6.33). Some representative results are given in Table 6.4. The authors allowed Z to be a fitting variable and obtained a best-fit value five times smaller than that of 10^{11} M^{-1} s^{-1} commonly assumed, but this is not a major influence on the fit because f_{AB} is near unity for most systems. Bernhard and Sargeson find that the measured self-exchange rate of 4 M^{-1} s^{-1} for the $Fe(OH_2)_6^{2/3+}$ system does not fit the results and treated this as a variable to obtain a self-consistent value of 6.2×10^{-3} M^{-1} s^{-1}. Deviations such as this were noted in the survey by Sutin et al. and have been examined more extensively by Hupp and Weaver,[27] who also suggest that a value of ~10^{-3} M^{-1} s^{-1} is more consistent for a number of reactions. The standard explanation for this deviation is that the $Fe(OH_2)_6^{2/3+}$ exchange is inner-sphere with a bridging water molecule. It has been suggested[28] that the self-exchange rate of 20 M^{-1} s^{-1} for the system $Ru(OH_2)_6^{2/3+}$ also implies a much smaller value for $Fe(OH_2)_6^{2/3+}$ because the inner-sphere rearrangement is greater for Fe than for Ru (see Table 6.1). In a similar vein, it should be noted that the difference in ionic radii of V(II) and V(III) is very similar to that of Fe(II) and Fe(III), so that the $Fe(OH_2)_6^{2/3+}$ exchange rate might be expected to be ~0.01 M^{-1} s^{-1}. On the other hand, the ΔV^* of -11 cm^3 mol^{-1} for the $Fe(OH_2)_6^{2/3+}$ exchange[29] has been taken as evidence for an outer-sphere

Table 6.4. Self-Exchange Rate Constants (M^{-1} s^{-1}, 25°C) Obtained by Fitting to the Marcus Cross Relationship

Reagent A	B	E_A (V)	E_B (V)	$k_{AA}{}^{a,b}$	$k_{BB}{}^{a,c}$	$10^{-4} \times k^a$ obsd	calc
$Ru(sar)^{2+}$	$(NH_3)_5Ru(py)^{3+}$	0.29	0.302	1.2×10^5	1.1×10^5	10.5	14
$Ru(sar)^{2+}$	$(NH_3)_5Ru(nic)^{3+}$	0.29	0.362	1.2×10^5	1.1×10^5	28	44
$Ru(sar)^{2+}$	$(NH_3)_5Ru(isn)^{3+}$	0.29	0.384	1.2×10^5	1.1×10^5	52	66
$Ru(sar)^{2+}$	$Ru(tacn)^{3+}$	0.29	0.366	1.2×10^5	5.4×10^4	73	34
$Ru(sar)^{2+}$	$Mn(sar)^{3+}$	0.29	0.519	1.2×10^5	1.7×10^1	17	9
$Ru(sar)^{2+}$	$Fe(OH_2)_6{}^{3+}$	0.29	0.74	1.2×10^5	6.2×10^{-3} ^{b}	7.2	6.7
$Mn(sar)^{3+}$	$(NH_3)_5Ru(py)^{2+}$	0.519	0.302	1.7×10^1	1.1×10^5	3.7	7.2
$Mn(sar)^{3+}$	$(NH_3)_5Ru(isn)^{2+}$	0.519	0.384	1.7×10^1	1.1×10^5	1.4	1.7
$Mn(sar)^{3+}$	$Ru(tacn)^{2+}$	0.519	0.366	1.7×10^1	1.2×10^5	2.9	1.6
$Mn(sar)^{3+}$	$Fe(OH_2)_6{}^{3+}$	0.519	0.74	1.7×10^1	6.2×10^{-3} ^{b}	1.2 ^{d}	2.0 ^{d}

^{a} $\mu = 0.1$ M, selected from Ref. 26, where structures of the ligands are given.
^{b} Determined from a least-squares fit to Eq. (6.33) with $Z = 1.9 \times 10^{10}$ M^{-1} s^{-1}.
^{c} Fixed values known independently unless otherwise indicated.
^{d} Values of $10^{-1} \times k$.

mechanism, especially by comparison to the value of 0.8 cm^3 mol^{-1} for $Fe(OH_2)_5OH^{2+}/Fe(OH_2)_6{}^{2+}$, which is believed to be inner-sphere. It is somewhat ironic that the mechanism for one of the most studied and analyzed reactions remains controversial.

The cross relationship can also be used to estimate self-exchange rates when these rates cannot be measured directly. If the least-squares analysis of Bernhard and Sargeson is not used, then the calculation is somewhat cyclical because k_{BB} also appears in f_{AB}, but this factor is often ~1 and not strongly dependent on k_{BB}. Macartney and Sutin[30] applied this method to various ascorbate radicals and their parents and calculated the following self-exchange rate constants (M^{-1} s^{-1}, 25°C): 2×10^3 for $H_2A/H_2A^{\bullet+}$; 1×10^5 for $HA/HA^\bullet$; and ~2×10^5 for $A^{2-}/A^{\bullet-}$.

Confidence in such applications is tempered by attempts to calculate the self-exchange rate constant for the dioxygen/superoxide (O_2/O_2^-) couple. Taube and co-workers[31] studied the oxidation of three Ru(II) ammine complexes and analyzed the results using Eq. (6.33) to obtain a reasonably self-consistent self-exchange rate constant of 1×10^3 M^{-1} s^{-1} for O_2/O_2^-. Espenson and co-workers[32] expanded the earlier study with more reducing agents and refined the analysis by including so-called work terms that attempt to correct for asymmetries in the charge and size of the species involved. The work term corrections are given by the following equations:[33]

$$k_{AB} = (k_{AA} k_{BB} K_{AB} f_{AB})^{1/2} W_{AB} \qquad (6.36)$$

where f_{AB} and W_{AB} are given by

$$\ln f_{AB} = \frac{(\ln K_{AB} + (w_{AB} - w_{BA})/RT)^2}{4 \ln (k_{AA} k_{BB}/Z^2) + (w_{AA} + w_{BB})/RT}$$

$$W_{AB} = \exp\left(\frac{-(w_{AB} + w_{BA} - w_{AA} - w_{BB})}{2RT}\right)$$

The individual work terms are calculated from

$$w_{ij} = \frac{4.225 \times 10^3 z_A z_B}{r (1 + 0.329 r \sqrt{\mu})} \qquad (6.37)$$

where r is the sum of the radii of the reaction partners in Å, μ is the ionic strength, the numerical constants are for water at 25°C, and the w_{ij} are in cal mol^{-1}. Some of these results are given in Table 6.5. The O_2 oxidations give values of k_{BB} which are consistent within the accepted limits, but the O_2^- reductions give quite different and divergent values. Finally, the self-exchange rate constant has been measured[34] by oxygen isotope exchange and found to be $(4.5 \pm 1.6) \times 10^2$ M^{-1} s^{-1} (in 0.3 M 2-propanol, 0.02 M NaOH). Several rationalizations for the discrepancy between the direct and Marcus relationship values may be offered. The direct O_2/O_2^- reaction may not be truly outer-sphere in that there might be some bonding interaction during the encounter of the reactants. In that case, the Marcus relationship might be giving the true outer-sphere rate constant by an argument analogous to that used for the $Fe(OH_2)_6^{2/3+}$ system. However, it may be that the Marcus formulation is not entirely justified for small and possibly strongly solvated species such as O_2^-. The small values obtained for the reductions by O_2^- present the unanswered problem of whether they represent the true outer-sphere rate constant or are due to some unexpected chemistry.

The results in Table 6.5 show that the corrections due to W_{AB} can be substantial because k_{BB} depends on W_{AB}^2. The small f_{AB} values for the O_2^- reactions result mainly from the large equilibrium constants for these reactions, which produce small values of f_{AB} because the denominator in the $\ln f_{AB}$ equation is negative.

This cross relationship is often applied to metalloenzyme systems to determine their self-exchange rates, because techniques are seldom available to measure the values directly. These applications have variable success, the difficulties usually being attributed to varying points of attack on the enzyme and induced conformational changes in the enzyme.

Table 6.5. O_2/O_2^- Self-Exchange Rate Constants[a] (25°C) as Calculated from the Marcus Cross Relationship

Reactants	k_{AB} (M^{-1} s^{-1})	$a_{ML}{}^b$ (Å)	W_{AB}	f_{AB}	$k_{BB}{}^c$ (M^{-1} s^{-1})
$Cr(bpy)_3{}^{2+} + O_2$	6×10^5	6.8	4.0	0.78	0.6
$Cr(phen)_3{}^{2+} + O_2$	1.5×10^6	6.8	4.0	0.72	1.9
$Ru(NH_3)_6{}^{2+} + O_2$	6.3×10^1	3.4	29.0	0.84	3.1
$Ru(en)_3{}^{2+} + O_2$	3.6×10^1	4.0	15.9	0.44	2.7×10^2
$Co(sep)^{2+} + O_2$	4.3×10^1	4.5	11.3	0.82	7.0×10^{-2}
$Fe(CN)_6{}^{3-} + O_2^-$	3×10^2	4.5	3.65	0.090	1.5×10^{-8}
$Fe(C_5H_5)_2{}^+ + O_2^-$	8.6×10^6	5.0	1.4	0.016	5.7×10^{-4}

[a] Further data and original references are given by Zahir, K.; Espenson, J. H.; Bakac, A. *J. Am. Chem. Soc.* **1988**, *110*, 5059.

[b] The radii of O_2 and O_2^- used are 1.21 and 1.33 Å, respectively, and the oxidized form of the metal complex was assumed to be 0.05 Å smaller than the reduced form, except for equal values for Ru complexes.

[c] $Z = 1 \times 10^{11}$ M^{-1} s^{-1} is used for all reactions.

6.3 DIFFERENTIATION OF INNER-SPHERE AND OUTER-SPHERE MECHANISMS

Several criteria can be used to differentiate the two mechanisms:

1. The best method is to identify a product to which the bridging group has been transferred, as in the classic study by Taube and co-workers discussed earlier. Unfortunately, the appropriate combination of substitution labile and inert metals is seldom available. Aside from Cr(II), other reagents that can be used are $Co(CN)_5{}^{3-}$ and bis(dimethylglyoxime)cobalt(II). Sometimes, both metal centers in the product are inert and the dimeric product can be identified as in the examples in Eqs. (6.38)[35] and (6.39):[36]

$$Fe(CN)_6{}^{3-} + Co(CN)_5{}^{3-} \longrightarrow (NC)_5FeCNCo(CN)_5{}^{6-} \quad (6.38)$$

$$IrCl_6{}^{2-} + Cr(OH_2)_6{}^{2+} \longrightarrow (Cl)_5IrClCr(OH_2)_5 + H_2O \quad (6.39)$$

2. If the rate constant for the oxidation–reduction reaction is larger than the rate of ligand substitution on either metal, then an outer-sphere mechanism is required. For example, $V(OH_2)_6{}^{2+}$ has a water exchange rate of 90 s^{-1} and substitution is by an I_a mechanism, so that the substitution rate constants should be <100 M^{-1} s^{-1}. For the following reactions, the rate constant is much larger than this:

$$(NH_3)_5CoOPO_3 + V(OH_2)_6{}^{2+} \xrightarrow{\quad k = 1.4 \times 10^7 \text{ M}^{-1} \text{ s}^{-1} \quad} \tag{6.40}$$

$$Fe(OH_2)_6{}^{3+} + V(OH_2)_6{}^{2+} \xrightarrow{\quad k = 1.8 \times 10^4 \text{ M}^{-1} \text{ s}^{-1} \quad} \tag{6.41}$$

In the last example, substitution might occur on Fe, but it has a water exchange rate of ~150 s^{-1} and the rate of the oxidation–reduction is also too fast for this pathway. Therefore, both reactions must be using an outer-sphere mechanism. Clearly, if both reactants are quite inert to substitution, an outer-sphere mechanism is almost certain.

3. If all the ligands on both reactants have no unshared electron pairs, it will not be possible to form a bridged intermediate. Ligands such as ammonia, bipyridyl, and *o*-phenanthroline are common examples. For the chelates, one must be certain that ring opening does not precede electron transfer for this criterion to be valid.

4. If the reaction(s) obey the predictions of the Marcus theory, then an outer-sphere mechanism is often assumed. This is the most dangerous criterion, because it has been observed that inner-sphere reaction rates[37,38] also show a correlation with the overall ΔG^o of the reaction that is predicted by the Marcus theory for outer-sphere reactions. Murdoch[39] has shown that such linear free-energy correlations may be more general than might originally have been expected.

6.4 BRIDGING LIGAND EFFECTS IN INNER-SPHERE REACTIONS

An inner-sphere mechanism consists of two processes, precursor complex formation followed by electron transfer:

$$L_5MX + NY_6 \underset{}{\overset{K}{\rightleftharpoons}} L_5M\!\!-\!\!X\!\!-\!\!NY_5 \xrightarrow{\ k_e\ } \text{Products} \tag{6.42}$$
$$+ Y$$

The experimental rate constant is a product of these two factors, as given by

$$k_{exp} = K\,k_e \tag{6.43}$$

and the variation of k_{exp} with the nature of the bridging ligand may reflect changes in either or both K and k_e. Rate constants for some inner-sphere reactions with different bridging ligands are given in Table 6.6.

Several kinetic trends have been interpreted as largely reflecting changes in K. When $(NH_3)_5CoX^{2+}$ complexes with X = F$^-$, Cl$^-$, and Br$^-$ are reduced by Cr^{2+}, the order of reactivity is Br$^-$ > Cl$^-$ > F$^-$, but the opposite order is observed when Eu^{2+} is the reducing agent. This is rationalized from the

Table 6.6. Rate Constants (25°C) for Reduction of $(NH_3)_5Co^{III}X$ Complexes by Cr^{2+} at $\mu = 1.0$ M

X	k $(M^{-1} s^{-1})$	ΔH^* $(kcal\ mol^{-1})$	ΔS^* $(cal\ mol^{-1}\ K^{-1})$
F^- [a]	2.5×10^5		
Cl^- [a]	6×10^5		
Br^- [b]	1.4×10^6		
HCO_2^- [c]	7.2	8.3	−27
$H_3CCO_2^-$ [c]	3.5×10^{-1}	8.2	−33
$Cl_2HCO_2^-$ [c]	7.5×10^{-2}	8.1	−36
$F_3CCO_2^-$ [c]	1.7×10^{-2}	9.3	−35
$(CH_3)_3CCO_2^-$ [c]	7.0×10^{-3}	11.1	−31
$HO_2CCO_2^-$ [d]	1.0×10^2		
$^-O_2CCO_2^-$ [d]	4.6×10^4	2.3	−20
$H_3CC(=O)CO_2^-$ [d]	1.1×10^4	5.8	−21
$H_3CC(OH)_2CO_2^-$ [e]	2.6×10^1		

[a] Candlin, J. P.; Halpern, J. *Inorg. Chem.* **1965**, *4*, 766.
[b] Moore, M. C.; Keller, R. N. *Inorg. Chem.* **1971**, *10*, 747; $\mu = 0.1$ M.
[c] Barrett, M. B.; Swinehart, J. H.; Taube, H. *Inorg. Chem.* **1971**, *10*, 1983.
[d] Price, H. J.; Taube, H. *Inorg. Chem.* **1968**, *7*, 1.
[e] Sisley, M. J.; Jordan, R. B. *Inorg. Chem.* **1989**, *28*, 2714.

knowledge that Eu^{2+} is a harder acid and forms stronger complexes with F^-, whereas Cr^{2+} forms stronger complexes with Br^-. If X is a carboxylate anion, then the order of reactivity of $(NH_3)_5CoX^{2+}$ with Cr^{2+} is $HCO_2^- > CH_3CO_2^- > CHCl_2CO_2^- > CF_3CO_2^- > (CH_3)_3CCO_2^-$. This is interpreted as a combination of steric and electron withdrawing effects causing K to decrease for the bridged intermediate. The unusually high reactivity of the formate complex has been attributed[40] to its ability to form the more sterically accessible conformer on the right in Eq. (6.44), therefore giving an even larger K than expected:

$$(6.44)$$

The much higher rates for the oxalato and keto form of pyruvato complexes[41] are attributed to stabilization of the bridged intermediate by chelation. The same effect may be operating, but less effectively, for the pyruvate hydrate complex.[42]

The kinetic product[43] shown in reaction (6.45) indicates that the simple carboxylate ions probably use the β-oxygen in bridging:

$$\left(\begin{array}{c} (H_3N)_5Co\!-\!S \diagup\!\!\!\!\diagdown O \\ \overset{|}{\underset{H}{\diagdown}} \overset{C}{\underset{N}{|}} \diagup CH_3 \end{array} \right)^{2+} \xrightarrow[11\ H_2O,\ 5\ H^+]{Cr^{2+}} \left(\begin{array}{c} S \diagup\!\!\!\!\diagdown O\!-\!Cr(OH_2)_5 \\ \overset{\diagup}{\underset{H}{\diagdown}} \overset{C}{\underset{N}{|}} CH_3 \end{array} \right)^{2+} \tag{6.45}$$

$$+\ Co(OH_2)_6{}^{2+}\ +\ 5\ NH_4{}^+$$

The O-bonded Cr(III) complex isomerizes to the stable S-bonded form. It was also shown that the O-bonded Co(III) complex produces the S-bonded Cr(III) product.

With Cr(II) as the reducing agent, the Cr(III) product provides evidence of an inner-sphere mechanism, but in a more general sense it would be useful to know the necessary properties for a bridging ligand. The minimum requirement is two electron pairs, one to bond to the oxidizing agent and the other to bond to the reducing agent. Jordan and Balahura[44] have suggested that in ligands more complex than the halides and hydroxide, two metal centers are unlikely to bond to the same atom and the two metals must be bonded to atoms that are part of a conjugated system. Then the electron may transfer through the π or π^* orbital of the conjugated system. The evidence for this was based on observations of the Cr(III) products from the Cr(II) reductions of the Co(III) complexes in Figure 6.2.

The observation that $(NH_3)_5CoOH_2{}^{3+}$ is reduced very slowly by Cr(II)[45] indicates that water is not an effective bridging ligand. However, OH$^-$ is an effective bridge in the same system.

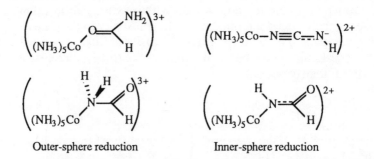

Outer-sphere reduction   Inner-sphere reduction

Figure 6.2. Cobalt(III) complexes whose chromium(II) reduction products indicate the reaction mechanism and the properties needed for a bridging ligand.

Inner-sphere attack of Cr(II) on the adjacent atom is observed with $(NH_3)_5Co$—SCN^{2+}, to yield $(H_2O)_5Cr$—SCN^{2+} ($k = 0.8 \times 10^5$ M^{-1} s^{-1}) as well as the remote attack product $(H_2O)_5Cr$—NCS^{2+} ($k = 1.9 \times 10^5$ M^{-1} s^{-1}).[46] The linkage isomer, $(NH_3)_5Co$—NCS^{2+}, gives only one product, $(H_2O)_5Cr$—SCN^{2+}. In the Co—NCS case, the adjacent N atom does not have a lone pair available to bond to the reducing agent, whereas the S atom does have a lone pair in the former.

The mechanistic classifications are not always as straightforward as the preceding examples might imply. Reduction of the N-bonded glycine complex[47] leads to transfer of glycine to Cr(III) although there is no conjugation between the $-NH_2$ and $-CO_2^-$ groups. This can be rationalized as due to a *bridged outer-sphere mechanism*, illustrated in Scheme 6.2.

Scheme 6.2

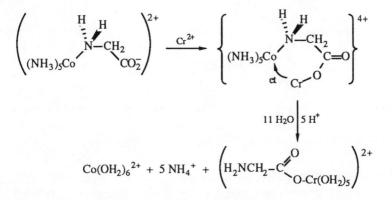

It is proposed that glycine serves to hold the oxidizing and reducing centers close together but that electron transfer proceeds by an outer-sphere process rather than through the glycine.

The systems listed in Table 6.7 are examples of what is called *remote attack*, in which the Cr(II) is attached at a ligand atom considerably removed from the Co(III). In such systems, the rate constant is sensitive to the nature of the remote substituent through its effect on K, but the rate constants also correlate generally with the ease of reduction of the bridging ligand, which seems to reflect an influence on k_e. The isonicotinamide system[48] gives an initial product with Cr(III) bonded to the oxygen of the amide group, and provides clear evidence for a remote attack mechanism. The meta isomer (nicotinamide) reacts about 500 times slower and gives about 70 percent Cr(III)-amide complex. The smaller rate of the *m*-acetylcyanobenzene complex compared to the *p*-acetyl isomer[49] also reflects the importance of conjugation between the remote group and the lead-in group at cobalt. These types of observations have led to the suggestion that such systems may proceed by a *chemical mechanism* in which the electron is actually transferred to the bridging group to form a radical intermediate.

Table 6.7. Rate Constants (25°C) for Reduction by Remote Attack of Cr^{2+}

X:Co(NH₃)₅	k (M⁻¹ s⁻¹)
H₂N–C(O)–C₆H₄–N:Co(NH₃)₅	17.4
H–C(O)–C₆H₄–CN:Co(NH₃)₅	2.5 x 10⁵
H₃C–C(O)–C₆H₄–CN:Co(NH₃)₅	6 x 10³
C₆H₄(CH₃CO)–CN:Co(NH₃)₅	0.28

In the case of *p*-formylbenzoato complex, the rate law has an H⁺ dependent path that can be rationalized by the mechanism in Scheme 6.3.[50]

Scheme 6.3

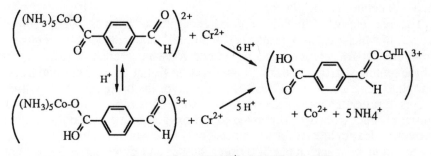

$$\text{Rate} = (\,5.3 + 380\,[\text{H}^+]\,)\,[\text{Co(III)}]\,[\text{Cr(II)}]$$

Protonation of the adjacent carboxylate group may increase the rate by improving the reducibility of the bridging ligand and/or improving the conjugation between the metal centers in the bridged intermediate. In analogous systems with simple carboxylate ligands, such as acetate and benzoate, protonation inhibits reduction,[51] presumably because the adjacent group cannot accommodate both H^+ and Cr^{2+}.

The radical mechanism has been characterized for a series of nitrobenzoate-type ligands attached to Co(III) in which moderately persistent radicals can be generated by pulse radiolysis.[52] The initial production of the coordinated radical uses a reactive radical (R•) such as e^-, $CO_2^{\bullet-}$, and $(H_3C)_2(C\bullet)OH$ produced in the radiolysis pulse, and the reaction proceeds as shown in Scheme 6.4. The rate of the intramolecular electron transfer (k_e) can then be measured, and some values of k_e are given in Table 6.8.

Scheme 6.4

$$(NH_3)_5Co\!-\!O_2C\!-\!X\!-\!C_6H_4NO_2{}^{2+} + R\bullet$$

$$\Big\downarrow \text{Fast}$$

$$\{(NH_3)_5Co\!-\!O_2C\!-\!X\!-\!C_6H_4\overset{\bullet}{N}O_2{}^+\} + R^+$$

$$\Big\downarrow k_e$$

$$^-O_2C\!-\!X\!-\!C_6H_4NO_2 + Co^{2+} + 5\,NH_3$$

It is notable that these intramolecular reactions are not exceptionally fast. The ortho isomers are much more reactive because of effective conjugation and possibly some "outer-sphere" transfer from the (•NO$_2$) group, which is near the Co(III). This is consistent with the lower rate when a CH=CH group is introduced. The meta isomers are least reactive because of poor conjugation. The moderate reactivity of the saturated $CH_2CH_2CH_2$ derivative is ascribed to the "outer-sphere" path allowed by the flexibility of the $CH_2CH_2CH_2$. The OC(NH)CH$_2$ derivative is less reactive because of the rigidity imposed by the planar OC(NH) group.

Table 6.8. Rate Constants (25°C) for the Reduction of Some Nitrobenzoate Radicals Coordinated to $(NH_3)_5Co^{III}$

X (isomer)	k_e (s^{-1})	X (isomer)	k_e (s^{-1})
- (o)	4.0×10^5	CH=CH (o)	1.7×10^3
- (m)	1.5×10^2	CH=CH (m)	3.1
- (p)	2.6×10^3	CH=CH (p)	4.8×10^2
CH$_2$ (o)	3.5×10^4	CH$_2$CH$_2$CH$_2$ (p)	1.5×10^2
CH$_2$ (m)	1.0×10^2	OC(NH)CH$_2$ (p)	5.8
CH$_2$ (p)	3.9×10^2		

Tsukahara and Wilkins[53] studied the product of the reaction of $CO_2^{\bullet-}$ with the 1-methyl-4,4'-bipyridinium (mbpy) complex $(NH_3)_5Co(mbpy)^{4+}$ at pH 7.2 and 25°C. The initial product was assigned as the radical complex $(NH_3)_5Co(mbpy^{\bullet})^{3+}$ which then undergoes intramolecular electron transfer $(k = 8.7 \times 10^2 \text{ s}^{-1})$ and bimolecular electron transfer to $(NH_3)_5Co(mbpy)^{4+}$ $(k = 5.4 \times 10^7 \text{ M}^{-1} \text{ s}^{-1})$. They also observed that the reaction of $(NH_3)_5Ru(mbpy)^{4+}$ and $CO_2^{\bullet-}$ produces no detectable radical intermediate and attribute this to fast intramolecular electron transfer $(k > 10^6 \text{ s}^{-1})$. The difference in reactivity of the Co and Ru systems was suggested to be due to the acceptor and donor orbitals both being of π symmetry in the Ru system, but the acceptor is of σ symmetry in the Co system.

Another potential method of separating the K and k_e effects on electron-transfer rates is actually to prepare the bridged complex using inert oxidizing and reducing centers. For example, Taube and co-workers[54] studied the following system:

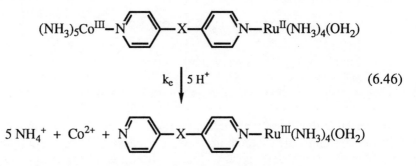

$$(6.46)$$

The values of k_e are rather insensitive to X (CH_2–CH_2, CH=CH) and small $(0.1 \times 10^{-2}$ to $4 \times 10^{-2} \text{ s}^{-1}$, 25°C), especially compared to 3 $\text{M}^{-1} \text{ s}^{-1}$ for the reaction of $(NH_3)_5Co(OH_2)^{3+}$ with $Ru(NH_3)_6^{2+}$. It was suggested that the reactions are slow because of inner-sphere reorganization analogous to that proposed for outer-sphere reactions of cobalt(III) complexes.

A similar study has been done on the analogous $(NH_3)_5Co$—L—$Fe(CN)_5$ system by Haim and co-workers.[55] The values of k_e are all in the range of 1.5×10^{-3} to $5 \times 10^{-3} \text{ s}^{-1}$, except for much smaller values when X = CH_2 or CO. The rate constants are similar to the values for the outer-sphere reactions of $(NH_3)_5Co(L')^{3+} + (CN)_5FeL^{3-}$ (where L and L' are pyridine derivatives). These reactions proceed through a strong ion pair ($K_{ip} \approx 900$ M^{-1}) so that k_e in the ion pair is measured.[56] The implication is that the binuclear systems with more flexible X, such as $(CH_2)_2$ and $(CH_2)_3$, use a bridged outer-sphere mechanism.

6.5 INTERVALENCE ELECTRON TRANSFER

There has been intense interest in the process referred to as intervalence electron transfer in mixed valence (oxidation state) species ever since the initial preparation[57] of the Taube–Creutz compound:

$$(H_3N)_5Ru^{II}—N\underset{}{\overset{}{\bigcirc}}N—Ru^{III}(NH_3)_5$$

The intervalence electron transfer involves electron exchange between the two metal centers that are in different oxidation states. This process seems to be typified by transitions in the near-infrared region of the electronic spectra of such species. For the preceding ion in water, this band occurs at 1560 nm (6,400 cm^{-1}). There is some debate as to whether the preceding description (localized or trapped valence) of this ion is correct or whether a delocalized picture is more appropriate. The trapped-valence model seems correct[58] for the complex in which the bridging ligand is 4,4'-bipyridine, since the two pyridine rings are not planar and therefore are not in full conjugation.

The unique electronic absorbance observed in these mixed-valence systems is usually assigned as a metal–metal charge transfer band (MMCT). Meyer and Hupp[59] have noted that it should not be a single band because of the lower than O_h symmetry and spin orbit coupling in Ru. The t_{2g} level will split into three nondegenerate levels so that three closely spaced bands may actually be observed. Since the MMCT process is equivalent to electron transfer from one metal to another, the interest in these systems has centered on the energetics of the MMCT process. Hush[60] suggested that the energy of the MMCT band would be the sum of the inner- and outer-sphere reorganization energies ($\Delta G_{in}^* + \Delta G_{solv}^*$) and it therefore presents a way of studying these features. Calculations by Creutz[58] indicate that ΔG_{in}^* is ~1400 cm^{-1} for Ru(II)/(III), so that ΔG_{solv}^* would seem to be the dominant factor given that the MMCT energies are in the 6000 to 12000 cm^{-1} range.

Hupp and Meyer measured the MMCT energy (E_{OP}) as a function of solvent for the 4,4'-bipyridine dimer in order to test the prediction of Hush that E_{OP} should be given by

$$E_{OP} = \Delta G_{in}^* + \frac{N e^2}{16 \pi \varepsilon_0}\left(\frac{1}{a} - \frac{1}{r}\right)\left(\frac{1}{n_s^2} - \frac{1}{\varepsilon_s}\right) \tag{6.47}$$

where it is assumed that $a_1 = a_2 = a$ in Eq. (6.25). The variation of E_{OP} with ($n_s^{-2} - \varepsilon_s^{-1}$) is linear, but the slope of 78,100 cm^{-1} is much larger than the predicted 22,500 cm^{-1} (a = 3.5 Å, r = 11.3 Å) and the intercept of 4820 cm^{-1} is rather far from the expected 1400 cm^{-1}. Meyer and Hupp have offered several explanations for these discrepancies, including the band multiplicity problem mentioned earlier.

The variation of E_{OP} with the structure of the bridging ligand has been the subject of several studies. It is hoped that this will provide some information on long-range electron-transfer reactions in biological systems.

Some results are given in Table 6.9 from studies by Sutton and Taube[61] and Spangler and co-workers (fifth entry)[62] on the pyridine derivatives, and by Stein et al.[63] on the thiospiranes. These examples fall into two categories. In the pyridine systems, the metal centers show significant coupling and the molar absorption coefficients (ε) are in the 100 to 1000 M^{-1} cm^{-1} range; in the thiospiranes, there is very weak coupling between the metals and the ε's are in the 2 to 50 M^{-1} cm^{-1} range. This difference means that electron transfer may be considered as being adiabatic for the pyridines, but probably is nonadiabatic for the thiospiranes. The substituents in the bridge clearly affect the coupling between the metal centers in a rational way that is reflected in the ε values. The two CH$_3$ substituents twist the pyridine rings further out of conjugation and reduce the coupling and the ε value, as does the saturated —CH$_2$— bridge. The conjugated —CH$_2$=CH$_2$— bridge allows conjugation between rings and increases ε. It should be noted that

Table 6.9. Intervalence Absorption Energies for Some Complexes of the Type (NH$_3$)$_5$Ru^{III}—L—Ru^{II}(NH$_3$)$_5$$^{5+}$

L	d (Å)	E$_{OP}$ (cm^{-1})	ε (M^{-1} cm^{-1})
	11.3	9,710	920
	11.3	11,240	165
	10.5	12,350	30
	13.8	10,420	760
	15.8	9,615	1430
	11.3	10,990	43
	14.4	12,300	9
	17.6	14,500	2.3

the data of Spangler and co-workers has recently been reanalyzed by Reimers and Hush[64] who have corrected the energies for band overlap. In the thiospiranes, the ε decreases as the Ru—Ru distance increases, indicating reduced coupling. The E_{OP} increases as the distance increases in the thiospiranes. The latter trends are expected for coupling through space or through σ bonds.

The nature of the relationship between intervalence electronic bands and electron-transfer processes remains an open question. There appears to be a relationship between E_{OP} and ΔG^*_{solv}, but the quantitative interpretation in terms of the Hush equation is less than satisfactory, as noted by Meyer and Hupp. Haim[65] has noted a correlation between the ΔG^* for electron transfer and the metal–metal separation for $(NH_3)_5Co$—L—$Fe(CN)_5$ and $(NH_3)_5Co$—L—$Ru(NH_3)_4(OH_2)$ systems. Geselowitz[66] found that in $(NH_3)_5Ru^{II}$—L—$M^{III}(NH_3)_5$ systems, E_{OP} for M = Ru correlates with ΔG^* for electron transfer for M = Co. He concluded that such systems have adiabatic electron transfer and the correlation works because E_{OP} is related to ΔG^*_{solv} when L gives reasonable coupling between metal centers.

The standard interpretation for weakly coupled systems has assumed that the inner- and outer-sphere rearrangement energies, sometimes called the nuclear factor, are independent of the separation between the metal centers and that the distance dependence of E_{OP} and ΔG^* for electron transfer is due to the decrease in electronic coupling between the centers with increasing distance. This electronic factor will affect the probability of electron transfer and therefore is lumped together with ΔS^* in transition-state theory interpretations. The dependence of the electronic factor (κ_e) on distance is taken to be related to that of the exchange integral (H_{AB}) between the metal centers, and quantum mechanics predicts that

$$\kappa_e = \kappa_{eo} \exp\left[-\beta(r - r_0)\right] \tag{6.48}$$

where $\beta \approx 1$ Å^{-1} (values of 0.9 to 1.2 are often used), r is the metal–metal separation in the system under study, and r_0 is the separation at which the transfer will be adiabatic (when $\kappa_e = \kappa_{eo}$). The value of r_0 is not known and is sometimes taken to be 0, although 3 to 4 Å might be more reasonable for metal ion complexes. The electron-transfer rate should have an exponential dependence on the metal–metal separation if other factors are constant.

Sutin and co-workers[67] have questioned the assumption that the nuclear factors ($\Delta G^*_{in} + \Delta G^*_{solv}$) are independent of the separation. These authors find a correlation between separation and E_{OP} for Ru^{II}—L—Ru^{III} systems and find that the ΔH^* for electron transfer in Os^{II}—$iso(Pro)_n$—Ru^{III} systems fits the same correlation, where $iso(Pro)_n$ is a polyproline bridge: with n = 1, r = 12.2 Å; n = 2, r = 14.8 Å; n = 3, r = 18.1 Å. This implies that the ΔH^* is dependent on distance and that it is not valid to assume that separation affects only the probability of electron transfer through the electronic factor.

6.6 ELECTRON TRANSFER IN METALLOPROTEINS

The metalloproteins consist of a metal complex imbedded in and bonded (through the ligand) to a protein net of covalently bonded amino acids. The most commonly studied systems are the myoglobins and cytochromes, which contain an iron (II or III) porphyrin complex, or the copper blue proteins, which have Cu(II) or Cu(I) complexed most often by histidine nitrogens and cysteine and methionine sulfurs from the protein. The metalloproteins can be oxidized or reduced by standard transition-metal complex reagents, and the latter usually are chosen to ensure outer-sphere electron transfer. This area has been the subject of several reviews.[68-73]

Since the metal center is surrounded by the protein system, these electron transfers occur at much longer distances (10–20 Å) than normal for small molecules and therefore have an analogy to the weakly coupled bimetal systems discussed in the previous section. The distance dependence of the rate of electron transfer has been investigated using modified proteins in which typically $(NH_3)_5Ru^{III}$ is attached at a specific site; then it is reduced to Ru(II) and the rate of electron transfer from the latter to the metal in the protein is measured.[74]

The interpretation of the kinetic results on these systems has revolved around the distance dependence, as discussed in the previous section. For example, a study[75] of Ru-modified myoglobins gave the electron-transfer rate as $k = 7.8 \times 10^8 \exp[\,0.91(r-3)\,]$ s^{-1}. The rate constants for these systems can be quite substantial and are often in the range of 1 to 10^3 s^{-1}.

Recent theoretical work indicates that such interpretations may be overly simplistic. The theory proposed by Hopfield and co-workers[76] suggests that the electron transfers through the σ bonds in the protein by a tunneling mechanism. The rate is then related to the most favorable electron-transfer path that can be found from a reagent at a particular site on the metalloprotein to its metal center. The theory has been applied recently[77] and found to be consistent with the relative rate constants observed for Ru-modified cytochrome c, but the simple-distance theory is also consistent with these observations. Beratan et al.[78] have surveyed several ruthenated proteins and devised methods for predicting the most favorable electron-transfer pathway. Ulstrup and co-workers[79] have given an extended Hückel theory analysis of the electron-transfer routes along different amino acid sequences in platocyanins.

Stein et al.[63] analyzed the ruthenium(II)(III)-thiospirane systems with an earlier version of Hopfield's theory and calculated electron-exchange rate constants of 8×10^7 and 3.5×10^4 s^{-1} for the two- and four-ring systems, respectively. A puzzling feature is why Taube and Haim did not see much larger rates in studies on the Co^{III}—L—Ru^{II} and Co^{III}—L—Fe^{II} systems. The problem may be with the large inner-sphere reorganization energy required by $(NH_3)_5Co^{III}$.

References

1. Taube, H.; Myers, H.; Rich, R. L. *J. Am. Chem. Soc.* **1953**, *75*, 4118; Taube, H.; Myers, H. *J. Am. Chem. Soc.* **1954**, *76*, 2103.
2. Marcus, R. A. *Annu. Rev. Phys. Chem.* **1964**, *15*, 155; *J. Chem. Phys.* **1965**, *43*, 679.
3. Hush, N. S. *Trans. Farad. Soc.* **1961**, *57*, 557; *Electrochim. Acta* **1968**, *13*, 1005; *Prog. Inorg. Chem.* **1967**, *8*, 391.
4. Levich, V. G. *Adv. Electrochem. Eng.* **1966**, *4*, 249.
5. Dogonadze, R. R. In *Reactions of Molecules at Electrodes*; Hush, N. S., Ed.; Wiley-Interscience: New York, 1971, Ch. 3.
6. Ratner, M. A.; Levine, R. D. *J. Am. Chem. Soc.* **1980**, *102*, 4898.
7. Sutin, N. *Acc. Chem. Res.* **1968**, *1*, 225.
8. Newton, T. W. *J. Chem. Educ.* **1968**, *45*, 571.
9. Sutin, N. *Prog. Inorg. Chem.* **1983**, *30*, 441.
10. Reynolds, W. L.; Lumry, R. W. *Mechanisms of Electron Transfer*; Ronald Press: New York, 1966.
11. Cannon, R. D. *Electron Transfer Reactions*; Butterworths: London, 1980.
12. Tembe, B. L.; Friedman, H. L.; Newton, M. J. *J. Chem. Phys.* **1982**, *76*, 1490.
13. Bernhard, P.; Ludi, A. *Inorg. Chem.* **1984**, *23*, 870.
14. Chan, M.-S.; Wahl, A. C. *J. Chem. Phys.* **1982**, *86*, 126.
15. Nielson, R. M.; McManis, G. E.; Safford, L. K.; Weaver, M. J. *J. Chem. Phys.* **1989**, *93*, 2152.
16. Lay, P. A.; McAlpine, N. S.; Hupp, J. T.; Weaver, M. J.; Sargeson, A. M. *Inorg. Chem.* **1990**, *29*, 4322.
17. Swaddle, T. W. *Inorg. Chem.* **1990**, *29*, 5017.
18. Binstead, R. A.; Beattie, J. K.; Dewey, T. G.; Turner, D. H. *J. Am. Chem. Soc.* **1980**, *102*, 6442.
19. Creaser, I. I.; Geue, R. J.; Harrowfield, J. McB.; Herlt, A. J.; Sargeson, A. M.; Snow, M. R.; Springborg, J. *J. Am. Chem. Soc.* **1982**, *104*, 6016.
20. Geselowitz, D. *Inorg. Chem.* **1981**, *20*, 4457.
21. Bernhard, P.; Sargeson, A. M. *Inorg. Chem.* **1988**, *27*, 2582.
22. Küppers, H.-J.; Neves, A.; Pomp, C.; Ventur, D.; Wieghardt, K.; Nuber, B.; Weiss, J. *Inorg. Chem.* **1986**, *25*, 2400.
23. Dubs, R. V.; Gahan, L. R.; Sargeson, A. M. *Inorg. Chem.* **1983**, *22*, 2523.
24. Dulz, G.; Sutin, N. *Inorg. Chem.* **1963**, *2*, 917; Ford-Smith, M. H.; Sutin, N. *J. Am. Chem. Soc.* **1961**, 83, 1830.
25. Chou, M.; Creutz, C.; Sutin, N. *J. Am. Chem. Soc.* **1977**, *99*, 5615.
26. Bernhard, P.; Sargeson, A. M. *Inorg. Chem.* **1987**, *26*, 4122.
27. Hupp, J. T.; Weaver, M. J. *Inorg. Chem.* **1983**, *22*, 2557.
28. Bernhard, P.; Helm, L.; Ludi, A.; Merbach, A. E. *J. Am. Chem. Soc.* **1985**, *107*, 312.
29. Jolley, W. H.; Stranks, D. R.; Swaddle, T. W. *Inorg. Chem.* **1990**, *29*, 1948.
30. Macartney, D. H.; Sutin, N. *Inorg. Chim. Acta* **1983**, *74*, 221.
31. Stanbury, D. M.; Haas, O.; Taube, H. *Inorg. Chem.* **1980**, *19*, 518.
32. Zahir, K.; Espenson, J. H.; Bakac, A. *J. Am. Chem. Soc.* **1988**, *110*, 5059.
33. Sutin, N. *Acc. Chem. Res.* **1982**, *15*, 275.
34. Lind, J.; Shen, X.; Merényi, G.; Jonsson, B. O. *J. Am. Chem. Soc.* **1989**, *111*, 7655.

35. Haim, A.; Wilmarth, W. K. *J. Am. Chem. Soc.* **1961**, *83*, 509.
36. Sykes, A. G.; Thorneley, R. N. F. *J. Chem. Soc. A* **1970**, 232.
37. Hua, L. H.-C.; Balahura, R. J.; Fanchiang, Y.-T.; Gould, E. S. *Inorg. Chem.* **1978**, *17*, 3692.
38. Linck, R. G. *Inorg. React. Methods* **1986**, *15*, 68.
39. Murdoch, J. R. *J. Am. Chem. Soc.* **1972**, *94*, 4410.
40. Balahura, R. J.; Jordan, R. B. *Inorg. Chem.* **1973**, *12*, 1438.
41. Price, H. J.; Taube, H. *Inorg. Chem.* **1968**, *7*, 1.
42. Sisley, M. J.; Jordan, R. B. *Inorg. Chem.* **1989**, *28*, 2714.
43. Balahura, R. J.; Johnson, M. D.; Black, T. *Inorg. Chem.* **1989**, *28*, 3933.
44. Jordan, R. B.; Balahura, R. J. *J. Am. Chem. Soc.* **1971**, *93*, 625.
45. Toppen, D. L.; Linck, R. G. *Inorg. Chem.* **1971**, *10*, 2635.
46. Shea, C.; Haim, A. *J. Am. Chem. Soc.* **1971**, *93*, 3055.
47. Kupferschmidt, W. C.; Jordan, R. B. *Inorg. Chem.* **1981**, *20*, 3469.
48. Nordmeyer, F.; Taube, H. *J. Am. Chem. Soc.* **1968**, *90*, 1162.
49. Balahura, R. J.; Purcell, W. L. *J. Am. Chem. Soc.* **1976**, *98*, 4457.
50. Zannella, A.; Taube, H. *J. Am. Chem. Soc.* **1972**, *94*, 6403.
51. Barrett, M. B.; Swinehart, J. H.; Taube, H. *Inorg. Chem.* **1971**, *10*, 1983.
52. Whitburn, K. D.; Hoffman, M. Z.; Simic, M. G.; Brezniak, N. V. *Inorg. Chem.* **1980**, *19*, 3180; Whitburn, K. D.; Hoffman, M. Z.; Brezniak, N. V.; Simic, M. G. *Inorg. Chem.* **1986**, *25*, 3037.
53. Tsukahara, K.; Wilkins, R. G. *Inorg. Chem.* **1989**, *28*, 1605.
54. Fischer, H.; Tom, G. M.; Taube, H. *J. Am. Chem. Soc.* **1976**, *98*, 5512.
55. Jwo, J.-J.; Gaus, P. L.; Haim, A. *J. Am. Chem. Soc.* **1979**, *101*, 6189.
56. Gaus, P. L.; Villanueva, J. L. *J. Am. Chem. Soc.* **1980**, *102*, 1934.
57. Creutz, C.; Taube, H. *J. Am. Chem. Soc.* **1969**, *91*, 3988; Ibid. **1973**, *95*, 1086.
58. Creutz, C. *Inorg. Chem.* **1978**, *17*, 3723.
59. Hupp, J. T.; Meyer, T. J. *Inorg. Chem.* **1987**, *26*, 2332.
60. Hush, N. S. *Inorg. Chem.* **1967**, *8*, 391.
61. Sutton, J. E.; Taube, H. *Inorg. Chem.* **1981**, *20*, 3125.
62. Woitellier, S.; Launay, J. P.; Spangler, C. W. *Inorg. Chem.* **1989**, *28*, 758.
63. Stein, C. A.; Lewis, N. A.; Seitz, G. *J. Am. Chem. Soc.* **1982**, *104*, 2596.
64. Reimers, J. R.; Hush, N. S. *Inorg. Chem.* **1990**, *29*, 4510.
65. Haim, A. *Pure Appl. Chem.* **1983**, *55*, 89.
66. Geselowitz, D. *Inorg. Chem.* **1987**, *26*, 4135.
67. Isied, S. S.; Vassilian, A.; Wishart, J. F.; Creutz, C.; Schwarz, H. A.; Sutin, N. *J. Am. Chem. Soc.* **1988**, *110*, 635.
68. Isied, S. S. *Prog. Inorg. Chem.* **1984**, *32*, 443.
69. Sykes, A. G. *Chem. Soc. Rev.* **1985**, *14*, 283.
70. McLendon,G.; Guarr, T.; McGuire, M.; Simolo, K.; Strauch, S.; Taylor, K. *Coord. Chem. Rev.* **1985**, *64*, 113.
71. Marcus, R. A.; Sutin, N. *Biochim. Biophys. Acta* **1985**, *811*, 265.
72. Gray, H. B. *Chem. Soc. Rev.* **1986**, *15*, 17.
73. McLendon, G. *Acc. Chem. Res.* **1988**, *21*, 160.
74. Scott, R. A.; Mauk, A. G.; Gray, H. B. *J. Chem. Educ.* **1985**, *62*, 932; Karas, J. L.; Lieber, C. M.; Gray, H. B. *J. Am. Chem. Soc.* **1988**, *110*, 599.
75. Axup, A. W.; Albin, M.; Mayo, S. L.; Crutchley, R. J.; Gray, H. B. *J. Am. Chem. Soc.* **1988**, *110*, 435.

76. Beratan, D. N.; Onuchic, J. N.; Hopfield, J. J. *J. Chem. Phys.* **1987**, *86*, 4488; Cowan, J. A.; Upmacis, R. K.; Beratan, D. N.; Onuchic, J. N.; Gray, H. B. *Ann. N. Y. Acad. Sci.* **1989**, *550*, 68.
77. Bowler, B. E.; Meade, T. J.; Mayo, S. L.; Richards, J. H.; Gray, H. B. *J. Am. Chem. Soc.* **1989**, *111*, 8757.
78. Beratan, D. N.; Onuchic, J. N.; Betts, J. N.; Bowler, B. E.; Gray, H. B. *J. Am. Chem. Soc.* **1990**, *112*, 7915.
79. Christensen, H. E. M.; Conrad, L. S.; Mikkelsen, K. V.; Nielsen, M. K.; Ulstrup, J. *Inorg. Chem.* **1990**, *29*, 2808.

7

Inorganic Photochemistry

Electromagnetic radiation in the form of ultraviolet and visible light has long been used as a reactant in inorganic reactions. The energy of light in the 200- to 800-nm region varies between 143 and 36 kcal mol^{-1}, so it is not surprising that chemical bonds can be affected when a system absorbs light in this readily accessible region. Systematic mechanistic studies in this area have benefited greatly from the development of lasers that provided intense monochromatic light sources, and from improvements in actinometers to measure the light intensity. Prior to the laser era, it was necessary to use filters to limit the energy of the light used to a moderately narrow region or to just cut off light below a certain wavelength. Pulsed-laser systems also allow much faster monitoring of the early stages of the reaction and the detection of primary photolysis intermediates.

The systems discussed in this chapter have been chosen because of their relationship to substitution reaction systems discussed previously. For a broader assessment of this area various books[1–5] and review articles[6–12] should be consulted.

7.1 BASIC TERMINOLOGY

Mechanistic photochemistry incorporates features of both electron transfer and substitution reactions, but the field has some of its own terminology, which is summarized as follows:

Quantum Yield (Φ)
The quantum yield is the number of molecules of reactant photolyzed divided by the number of quanta of light absorbed. An einstein (E) is defined as a mole of quanta, and if n is the moles of reactant photolyzed, then $\Phi = n/E$. The quantum yield also may be defined in terms of the moles of product formed.

Actinometer
An actinometer is a device used to measure the number of einsteins emitted at a particular wavelength by a particular light source. Photon-counting devices are now available and secondary chemical actinometers have been

developed, such as that based on the Reineckate ion[13] $(Cr(NH_3)_2(NCS)_4^-$, as well as the traditional iron(III)–oxalate and uranyl–oxalate actinometers. An early problem in this field was the lack of an actinometer covering the 450- to 600-nm range and the Reineckate actinometer solved this problem.

Reactant Photolyzed (n)
The reactant photolyzed is determined by appropriate analytical techniques. A combination of spectrophotometry and chromatography is commonly used. This is not a trivial problem because photochemical studies typically follow only the first 5 to 15 percent of the reaction, so that small amounts of product must be determined in the presence of large amounts of reactant.

Internal Filtration
If the reaction products absorb light at the wavelength being used, then the quantum yield will decrease as the reaction proceeds because reactants are not absorbing all the light. To minimize this problem, only the initial stages of the photochemical process are studied.

Secondary Photolysis
Secondary photolysis refers to the photolysis of the initial products to give secondary products. Again, only the initial part of the primary reaction is followed to minimize this problem.

Stern–Volmer Plots
Stern–Volmer plots are used to test the dependence of the quantum yield on the concentration of reactants. The form of the plot depends on the photochemical mechanism proposed.

Fluorescence
Fluorescence refers to the emission of light when an electronic excited state decays to another state of the same spin multiplicity. The emission is usually very fast.

Phosphorescence
Phosphorescence refers to the emission of light when an electronic excited state decays to another state of different spin multiplicity. This process is usually slower than fluorescence and is typically on the millisecond to microsecond time scale for transition-metal complexes.

Sensitizer
A sensitizer is a substance that makes a reaction more sensitive to photolysis. Sensitizers absorb light more strongly than the reactant and then transfer the absorbed energy to the reactant. The sensitizer must have an electronic excited state that is sufficiently long-lived to allow for the energy transfer to the reactant, and the energy of this state must be similar to that of the acceptor state of the reactant.

Quencher

A quencher is a substance that reduces the quantum yield of a process by accepting energy from the photoexcited state(s) of the reactant. The energy of the excited state of the quencher must be similar to that of the reactant.

Photostationary State

A photostationary state can occur in a kinetically labile system, initially at equilibrium, in which only the forward reaction is promoted by light. Under photochemical conditions, more reactant will be converted to product and a new equilibrium condition will be established in which the forward and reverse rates are the same; this is referred to as a photostationary state. If the light source is removed, the system will return to the thermodynamic equilibrium position.

Intersystem Crossing

Intersystem crossing refers to the process whereby an electronic excited state may be converted to another excited state of similar or lower energy. Back intersystem crossing refers to the reverse process.

7.2 KINETIC FACTORS AFFECTING QUANTUM YIELDS

Most photochemical systems have some common features that affect the lifetime of the photoexcited states and thereby the quantum yields. Figure 7.1 describes a general system with a ground state G which absorbs a photon at a rate (dE/dt) to produce an excited state I. The latter can undergo intersystem crossing to produce the photoactive state A which decays to products P or back to the ground state. Excited vibrational levels within each electronic state have been omitted for simplicity.

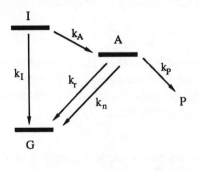

k_I non radiative deactivation

k_A conversion of I to A

k_r radiative deactivation of A

k_n non radiative deactivation of A

k_P production of product P

$$\Phi = \frac{dP/dt}{dE/dt} = \frac{dP}{dE}$$

Figure 7.1. A general photochemical system with a ground state G, an initially activated state I, and a photoactive state A.

It is normal to assume a steady state for the excited states A and I. Then

$$[A] = \frac{k_A[I]}{(k_P + k_r + k_n)} \tag{7.1}$$

$$[I] = \frac{1}{k_A + k_I}\left(\frac{dE}{dt}\right) \tag{7.2}$$

so that

$$[A] = \frac{k_A}{(k_A + k_I)(k_P + k_r + k_n)}\left(\frac{dE}{dt}\right) \tag{7.3}$$

Since

$$\frac{d[P]}{dt} = k_P[A] \tag{7.4}$$

then

$$\frac{d[P]}{dt} = \frac{k_P k_A}{(k_A + k_I)(k_P + k_r + k_n)}\left(\frac{dE}{dt}\right) \tag{7.5}$$

and the quantum yield is given by

$$\Phi = \frac{d[P]}{dt}\left(\frac{dE}{dt}\right)^{-1} = \frac{k_P k_A}{(k_A + k_I)(k_P + k_r + k_n)} \tag{7.6}$$

The point of this development is to show that several factors in addition to k_P can affect the quantum yield. It is not uncommon in this area to study the effect of changing ligand substituents and solvents on Φ and use the results to infer the mechanism of the k_P step. But such changes in conditions may affect k_A, k_I, k_r, and/or k_n and thereby make any mechanistic conclusions very tenuous. Of course, the system in Figure 7.1 could be expanded to include formation of products from the initially populated state I or from other photoactive states produced from I or A.

7.3 PHOTOCHEMISTRY OF COBALT(III) COMPLEXES

Cobalt(III) forms a wide range of substitution inert (low-spin d^6) complexes whose thermal aquation and anation reactions have been thoroughly studied. These provide useful comparisons for photochemical work. In addition, the substitution inertness of the products is an advantage for product studies.

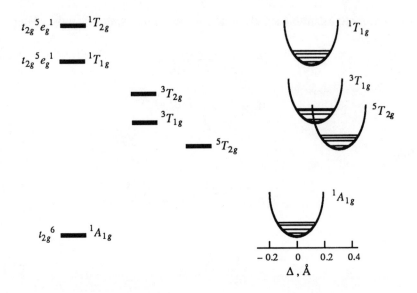

Figure 7.2. The ligand field electronic state energies and potential energy surfaces for octahedral cobalt(III).

7.3.a Co(III)L$_6$ Complexes

The electronic spectroscopy of Co(III)L$_6$ systems is well understood and described by ligand field theory. The electronic states involving the d orbitals are shown in Figure 7.2. The electronic spectra generally show two absorptions in the visible region, $^1A_{1g} \to {}^1T_{1g}$ (~500 nm) and $^1A_{1g} \to$ $^1T_{2g}$ (~350 nm). The spin-forbidden transitions to the $^3T_{1g}$ and $^5T_{2g}$ states are normally too weak to be observed. In addition, there often is a ligand-to-metal charge-transfer band in the ultraviolet region.

The diagram at the right in Figure 7.2 shows the potential energy surfaces for four of these levels in Co(NH$_3$)$_6^{3+}$, as suggested by the low-temperature spectroscopic study of Wilson and Solomon.[14] The Co—N bond lengths change by ΔÅ from the ground-state values.

The photoaquation of Co(NH$_3$)$_6^{3+}$ has a very low quantum yield of 3.1×10^{-4} mol einstein^{-1} for wavelengths in the visible region. The photoactive state is usually assumed to be the $^3T_{1g}$, which is populated by intersystem crossing from the $^1T_{1g}$ and $^1T_{2g}$ states. The low quantum yield could be associated with radiationless deactivation of these states. Wilson and Solomon suggest that the $^3T_{1g}$ may decay to the $^5T_{2g}$ state, which then decays efficiently to the ground state because of the overlap of the potential energy surfaces.

According to Scandola et al.[15] and Nishazawa and Ford,[16] the photoaquation of Co(CN)$_6^{3-}$ is much more efficient ($\Phi = 0.31$) for wavelengths in the visible region. Wilson and Solomon have suggested that the $^5T_{2g}$ state is at higher energy than the $^3T_{1g}$ state in this system because of the larger Dq of CN^- compared to NH_3; therefore, deactivation through the quintet state is not effective in $Co(CN)_6^{3-}$.

Scandola et al. studied the solvent dependence of the quantum yield in water–glycerol solutions and found that Φ decreases from 0.31 to 0.1 with increasing amounts of glycerol. They interpreted this as a viscosity effect on a cage intermediate, shown in Scheme 7.1.

Scheme 7.1

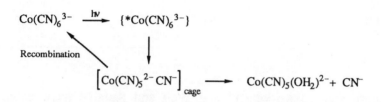

It was suggested that increasing viscosity inhibited CN^- release from the cage and therefore favored recombination to give a lower quantum yield. This type of explanation is not uncommon in photochemical studies, but it is now recognized that solvent changes can also affect the lifetimes of the photoactive states by changing the efficiency of the various decay mechanisms, and meaningful interpretations require information about excited-state lifetimes in the solvent mixtures. In the case of $Co(CN)_6^{3-}$, Wong and Kirk[17] found that $Co(CN)_5(glycerol)^{2-}$ is actually formed.

7.3.b Co(III)L$_5$Y Complexes

$Co^{III}(CN)_5X$ complexes photoaquate cleanly to $Co^{III}(CN)_5(OH_2)^{2-}$ with quantum yields in the range of 0.05 to 0.3, depending on the nature of X. Kirk and Kneeland[18] used competition with SCN^- to compare the thermal and photochemical reactions of $Co^{III}(CN)_5X^{3-}$ complexes. The ratio of the linkage isomers $Co^{III}(CN)_5(SCN)^{3-}$ and $Co^{III}(CN)_5(NCS)^{3-}$ is independent of the leaving group X but the ratio is different for the thermal and photochemical reactions, as shown in Table 7.1. The conclusion is that the transition states for the two processes are different because they give different isomer ratios, but both appear to be dissociative because the ratio is reasonably independent of the leaving group. Kirk and Kneeland suggest that the photoactive state, possibly $^3T_{1g}$, has a more diffuse electronic distribution and is therefore "softer," favoring bonding to the "softer" S end of thiocyanate.

The photolysis of $Co^{III}(NH_3)_5X$ complexes can be complicated by a photoredox process in which Co(III) is reduced to Co(II) and a ligand is oxidized. This is the dominant process when absorption is into the ligand-to-metal charge-transfer band ($\leq$350 nm) and it has quantum yields in the 0.1 to 0.5 range, depending on X. This process was studied extensively by

Table 7.1. Comparison of the S- and N- Linkage Isomers from Thermal and Photochemical Aquation of $Co(CN)_5X^{3-}$ in the Presence of SCN^-

Ligand X	Thermal S/N Ratio	Photochemical S/N Ratio
Cl^-	4.3	8.5
Br^-	4.5	8.8
I^-	4.7	9.3
N_3^-	4	8.7
CN^-		8.1
OH^-		8.7

Endicott and co-workers,[19] and Weit and Kutal[20] have examined $Co^{III}(NH_2CH_3)_5X$ systems with $X = Cl^-$ and Br^-. The NH_3 and NH_2CH_3 systems are more different than might be expected from the rather simple addition of a methyl group. The overall conclusions for these systems are summarized in Scheme 7.2.

Scheme 7.2

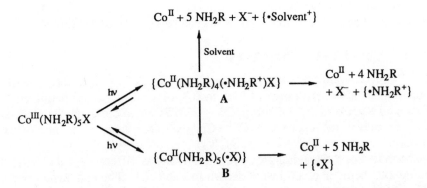

The electronic spectral properties indicate that species **A** will be favored by higher energy irradiation compared to species **B**. The NH_3 systems appear to undergo internal conversion to **B** or react with solvent, based on the solvent sensitivity of the reaction, and do not form any amine-type radicals. The NH_2R systems are different in that they show a wavelength dependence of the quantum yield, a different solvent variation in glycerol–water, and an increase in quantum yield in the presence of O_2. Weit and Kutal interpret these differences as due to shielding of the complex from the solvent by the CH_3 group so that reaction from **A** is observed. The O_2 effect is ascribed to scavenging of the •NH_2R radicals by O_2.

Table 7.2. Quantum Yields for the Loss of X and NH_3 for Complexes of the Type $Co^{III}(NH_3)_5X$

X	Φ_X	Φ_{NH_3}
F[a]	5.5×10^{-4}	2.0×10^{-3}
Cl[a]	1.7×10^{-3}	5.1×10^{-3}
Br[b]	2.0×10^{-3}	5.1×10^{-4}

[a] Zribush, R. A.; Poon, C. K.; Bruce, C. M.; Adamson, A. W. *J. Am. Chem. Soc.* **1974**, *96*, 3027, 25°C, pH 2, 488 nm; Langford, C. H.; Malkhasian, Y. S. *J. Am. Chem. Soc.* **1987**, *109*, 2682 give λ dependence with X = Cl⁻.
[b] Zanella, A. W.; Ford, K. H.; Ford, P. C. *Inorg. Chem.* **1978**, *17*, 1051, 25°C, pH 3.4, 546 nm.

If the irradiation is into the *d–d* bands (wavelength ≥450 nm), then photoaquation is observed with low quantum yields in the range of 1×10^{-2} to 1×10^{-3}. These reactions show a significant *antithermal pathway* in which NH_3 is released to yield $(NH_3)_4Co(X)(OH_2)$ mainly as the trans isomer. Quantum yields for the two paths are given in Table 7.2.

7.4 PHOTOCHEMISTRY OF RHODIUM(III) COMPLEXES

Rhodium(III) systems are formally analogous to cobalt(III) in that both are d^6 systems. The larger Dq values for second-row transition metals makes all the Rh(III) complexes more analogous to the cyano complexes of Co(III). Studies with rhodium(III) have the advantage that the rate of decay of the photoexcited states can be measured. The photoactive state appears to be the $^3T_{1g}$ (^3E for A_5RhX symmetry), and the rate constants for phosphorescent decay from this state in the solids at 77 K are given in Table 7.3.[21] The dramatic effect of changing hydrogen for deuterium in the hexaammine complex indicates that the energy of the excited state is dissipated to ligand

Table 7.3. Rate Constants (77 K) for Decay of the Triplet State for Some Rhodium(III) Complexes

Complex	k_n (s⁻¹)
$(NH_3)_5Rh(OH_2)^{3+}$	2.9×10^5
$(NH_3)_5RhCl^{2+}$	8.2×10^4
$(NH_3)_5RhNH_3^{3+}$	5.1×10^4
$(ND_3)_5RhND_3^{3+}$	0.8×10^4

Table 7.4. Photolysis (25°C) of $(NH_3)_5RhCl^{2+}$ in Various Solvents

Solvent	Φ_{Cl^-}	Φ_{NH_3}	k_n (s^{-1})
Water	0.18	0.02	7.0×10^7
Formamide	0.057	<0.011	4.5×10^7
Dimethylsulfoxide	<0.006	0.029	2.8×10^7
Methanol	0.008	0.11	$<5 \times 10^7$
Dimethylformamide	0.004	0.070	3.1×10^7

vibrational modes. This is probably a common feature for many complexes.

Solvent effects have been investigated[22] for the photolysis of $(NH_3)_5RhCl^{2+}$. The reaction proceeds with loss of either NH_3 or Cl^- in proportions that vary with the solvent, as shown by the quantum yields given in Table 7.4. The quantum yields do not correlate with the lifetime of the excited state, and the chloride loss process is especially sensitive to the solvent. A complicating feature may be the increasing recombination with Cl^- in the solvent cage when solvation of the chloride ion is less favorable.

The volumes of activation for the photoaquation of $(NH_3)_5RhCl^{2+}$ have been determined[23] as -8.6 and 9.3 cm^3 mol^{-1} for production of $(NH_3)_5RhOH_2^{3+}$ and *trans*-$(NH_3)_4Rh(OH_2)Cl^{2+}$, respectively. The large difference in these values implies that leaving group solvation is an important factor, but the analysis is complicated by the uncertain volumes of the electronic excited state(s). The photoaquation of *cis*-$Rh(bpy)_2Cl_2^+$ to *cis*-$Rh(bpy)_2Cl(H_2O)^{2+}$ has $\Delta V^* = -9.7$ cm^3 mol^{-1}. These two negative activation volumes for Cl^- release pose a problem for rationalizations[24] in terms of a dissociative mechanism.

Irradiation of the ligand-to-metal charge-transfer bands of $Rh(NH_3)_5I^{2+}$ gives *trans*-$Rh(NH_3)_4(H_2O)I^{2+}$ with a quantum yield about half of that for irradiation of the lower energy d–d bands.[25] This indicates that deactivation of the charge-transfer states is very competitive with intersystem crossing to the lower energy states. Flash photolysis studies in the presence of traces of I^- reveal the presence of transient $\cdot I_2^-$ and the redox mechanism in Scheme 7.3 has been proposed.

Scheme 7.3

$$Rh(NH_3)_5I^{2+} \xrightarrow[\text{Charge transfer}]{h\nu} Rh(NH_3)_4^{2+} + NH_3 + \cdot I$$

$$I^- + \cdot I \longrightarrow \cdot I_2^-$$

$$Rh(NH_3)_4^{2+} + \cdot I_2^- + H_2O \longrightarrow \textit{trans}\text{-}Rh(NH_3)_4(H_2O)I^{2+} + I^-$$

This behavior is quite different from that of the cobalt complexes in the preceding section, possibly because Rh(II) remains in the low-spin state, while Co(II) goes to the labile high-spin state.

7.5 PHOTOCHEMISTRY OF CHROMIUM(III) COMPLEXES

Chromium(III) systems have long been the prototype for inorganic photochemical studies. They have the same chemical advantages as cobalt(III) systems, but the quantum yields are much larger, the bands in the electronic spectra are well separated, and there is no photoredox process.

Some general observations can be made:
1. Quantum yields are typically 0.1 to 0.8.
2. The quantum yield is independent of the incident light energy.
3. The reactions often appear to be antithermal.

For example, photolysis[26] of $(NH_3)_5CrNCS^{2+}$ at 373 nm gives $\Phi_{NH_3} = 0.46$ and $\Phi_{NCS} = 0.03$, and at 492 nm the values are 0.47 and 0.021, respectively. Ammonia loss is the dominant photochemical process, but it is not observed thermally.

7.5.a Cr(III)L₆ Complexes

The electronic states involving the valence d orbitals in $Cr^{III}L_6$ complexes are shown in Figure 7.3. Electronic transitions to the $^4T_{2g}$ and $^4T_{1g}$ states are observed at 550 to 600 nm and 350 to 400 nm, respectively. Weak-spin-forbidden transitions to the $^2T_{1g}$ and ^2E_g states can sometimes be observed in the 600-nm region.

There has long been a controversy about the photoactive state(s) in these complexes, primarily concerning the $^4T_{2g}$ and doublet states ($^2T_{1g}$, ^2E_g). Some evidence favoring the quartet state was consistent with the idea that population of the e_g orbital would promote a dissociative mechanism. In the

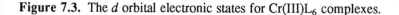

Figure 7.3. The d orbital electronic states for Cr(III)L₆ complexes.

doublet states, the metal ligand bonds would not be weakened much compared to the ground state, but the empty t_{2g} orbital could favor an associative substitution mechanism.

Waltz and Lillie[27] have used pulsed-laser techniques to measure the phosphorescent lifetimes of $Cr(NH_3)_6^{3+}$ and conductivity to monitor the following photoaquation reaction:

$$Cr(NH_3)_6^{3+} \xrightarrow[H_2O, H^+]{h\nu} Cr(NH_3)_5(OH_2)^{3+} + NH_4^+ \qquad (7.7)$$

They observed an initial fast decrease in conductivity ($t_{1/2} \approx 1 \times 10^{-6}$ s) followed by a slower change ($t_{1/2} \approx 10 \times 10^{-6}$ s). The slower process is the major one (>67 percent) and has the same rate as the phosphorescent decay, which is from a doublet state. They concluded that the fast conductivity change is due to photoaquation from the quartet state and the slower and dominant decay is from the doublet state. The general implication of these observations is that both doublet and quartet states can be photoactive, with their relative amounts depending on the efficiency of the intersystem crossing between these states, which in turn will depend on the ligands on the chromium(III).

For $Cr(CN)_6^{3-}$, Wasgestian[28] found that the emission lifetime of the doublet state(s) is very solvent dependent in N,N-dimethylformamide–water mixtures, but the quantum yield for aquation (0.11) is unaffected by the solvent composition. This shows that reactivity is not from the doublet state(s) and suggests that the quartet state is photoactive. From the activation volume[29] for photoaquation of 2.7 cm^{-3} mol^{-1} (15°C, 364.5 nm), an I_d mechanism has been suggested. In the related complex $Cr(CN)_5(NH_3)^{2-}$, quartet state activity also is indicated,[30] since $Co(sep)^{3+}$ quenches the phosphorescence in DMSO but does not change the quantum yields.

7.5.b   Cr(III) Amine Complexes

There has been a great deal of work on the photoaquation of complexes of the general type CrA_5X, CrA_4X_2, CrA_3X_3 and their isomers, where A is an amine ligand. The reactions are often, but not always, antithermal, showing loss of the amine ligand. The first attempt to explain the variation in products and quantum yields was made by Adamson.[31] These led to what are termed *Adamson's rules*, which can be summarized as follows:

1. The axis with the lowest average 10 Dq will be the most labilized.
2. If two different ligands are on that axis, the one with the larger Dq will photoaquate.
3. The quantum yield will be about the same as that for the CrL_6 complex, where L are the lower Dq axis ligands.

Some complexes that obey Adamson's rules are given by the first four examples in Table 7.5.

Table 7.5. Quantum Yields for Photoaquation of NH_3 (Φ_N) and X (Φ_X) from Some Chromium(III) Amine Complexes[a]

Complex	Φ_N	Φ_X
$Cr(NH_3)_5Cl^{2+}$	0.36	0.005
trans-$Cr(en)_2Cl_2^+$	<0.001	0.32
trans-$Cr(NH_3)_4Cl_2^+$	0.003	0.44
trans-$Cr(en)_2(NH_3)Cl^{2+}$	0.34	<0.01
trans-$Cr(en)_2F_2^+$ [b]	0.20	0.02
trans-$Cr(en)_2(NH_3)F^{2+}$	0.27	0.14

[a] Unless otherwise indicated, original references are given by Kirk, A. D. *Coord. Chem. Rev.* **1981**, *39*, 225.

[b] Manfrin, M. F.; Sandrini, A.; Juris, A.; Gandolfi, M. T. *Inorg. Chem.* **1978**, *17*, 90.

The fluoride complexes do not obey the "rules" because the Dq of F^- is smaller than that of the amine nitrogen. This led to considerable soul searching and the rationalization that what is really required is a measure of the Cr—X bond strength. With F^-, it was suggested that the Dq was anomalously low because of π-bonding effects, so that Dq did not reflect the σ-bond strength. This resulted in ligand field theory rationalizations such as those proposed by Zinck[32] and Vanquickenborne and Cuelemans.[33] The observation[34] that *trans*-(1,3-propanediamine)$_2Cr(F)_2^+$ gives mainly F^- loss ($\Phi_F = 0.34$, $\Phi_N = 0.18$) and is much different from the ethylenediamine analogue shows that subtle effects (steric or ring strain) can influence the course of the photochemical process.

Recent work on the competitive water exchange and NH_3 aquation[35] is also not consistent with the theoretical approaches. Some of these results are given in Table 7.6. The first entry is consistent with Adamson's and other

Table 7.6. Quantum Yields for the Water Exchange (Φ_{exch}) and NH_3 Loss (Φ_{NH_3}) from Some Chromium(III) Complexes

Complex	Φ_{exch}	Φ_{NH_3}
$Cr(NH_3)_5OH_2^{3+}$	0.078	0.195
cis-$Cr(NH_3)_4(OH_2)_2^{3+}$	0.057	0.058
trans-$Cr(NH_3)_4(OH_2)_2^{3+}$	0.001	0.025
fac-$Cr(NH_3)_3(OH_2)_3^{3+}$	0.040	0.053
trans-$Cr(NH_3)_2(OH_2)_4^{3+}$	0.072	0.004

predictions, but the second should predominantly give NH_3 loss, and the third and fourth should mainly give water exchange; the last entry conforms to the "rules."

The antithermal nature of the photochemical reactions has been reassessed recently.[36] The thermal reaction of $Cr(NH_3)_5OH_2^{3+}$ at 50°C is mainly water exchange (k = 1.37×10^{-3} s^{-1}, $\Delta H^* = 99.1$ kJ mol^{-1}) but a minor path gives cis-$Cr(NH_3)_4(OH_2)_2^{3+}$ (k $= 4.03 \times 10^{-6}$ s^{-1}, $\Delta H^* = 110.5$ kJ mol^{-1}). If one happened to work at a high enough temperature, then the ammonia loss could be the dominant thermal reaction. It was suggested that the photochemical process provides access, in the electronic ground state, to a pentagonal bipyramid transition state that is about 10 kJ mol^{-1} higher in energy than the transition state for thermal water exchange.

Some evidence for stereomobility of the photochemical transition state has been obtained[37] with the rigid macrocyclic system $trans$-$Cr(cyclam)(Cl)_2^+$. The quantum yield for Cl$^-$ aquation is very low (3×10^{-4}) and this has been attributed to the macrocycle preventing rearrangements that normally lead to less energetic configurations in the activated states which lead to aquation. The $trans$-$Cr(cyclam)(NH_3)_2^+$ is similar, but cis-$Cr(cyclam)(NH_3)_2^+$ has $\Phi_{NH_3} = 0.2$,[38] although the trans isomer has a much longer lived doublet state ($\tau = 55 \times 10^{-6}$ s, versus 2×10^{-6} s at 293 K). The authors propose that the photoactive state is the quartet state, which may be reached by back intersystem crossing. The latter process is less effective with the trans isomer because the energy separation is 22.2 kcal mol^{-1} compared to 19.0 kcal mol^{-1} for the cis isomer.

Volumes of activation have been measured by Angermann et al.[39] for some of the reactions and the results are given in Table 7.7. The authors suggest that the negative values of ΔV^* are consistent with associative activation for the photochemical process and the thermal reaction. Solvent electrostriction effects should make a larger negative contribution when X$^-$ is the leaving group compared to NH_3. Since this difference is not reflected in the ΔV^* values, Angermann et al. proposed that the ammonia release process is "more associative" than the X$^-$ release.

Table 7.7. Volumes of Activation for Photochemical and Thermal Aquation of Some Chromium(III) Complexes

	ΔV^* (cm^3 mol^{-1})		
Complex	Φ_{NH_3}	Φ_{X^-}	Thermal
$Cr(NH_3)_5Cl^{2+}$	−9.4	−13.0	−10.8
$Cr(NH_3)_5Br^{2+}$	−10.2	−12.2	−10.2
$Cr(NH_3)_5NCS^{2+}$	−11.4	−9.8	−8.6
$Cr(NH_3)_6^{3+}$	−12.6		

7.6 RU(II)POLYPYRIDINE COMPLEXES

Much of the work in the Ru(II)polypyridine area has concentrated on $Ru(bpy)_3^{2+}$ and its derivatives. The electronic spectra are dominated by an intense charge-transfer band in the visible region ($\lambda_{max} \approx 450$ nm, $\varepsilon_{max} \approx 10^4$ M^{-1} cm^{-1}). $Ru(bpy)_3^{2+}$ shows modest photosubstitution activity[40] with quantum yields <0.1, and photochemical methods have proved useful in the preparation of several derivatives,[41] including the unusual *trans*-$Ru(bpy)_2(OH_2)_2^{2+}$.[42]

$Ru(bpy)_3^{2+}$ is of interest because of its use in photochemical energy transfer, and its photophysics and applications have been the subject of several recent reviews.[43–46] Parris and Brandt[47] first observed that this complex has a relatively long-lived emission ($\tau \approx 600$ ns in water). This, together with its electronic spectral properties, means that the system can trap visible light energy long enough for the photoexcited state to undergo further chemical reactions. The long-lived excited state is a metal-to-ligand charge-transfer triplet which can act as a reducing agent to produce $Ru(bpy)_3^{3+}$ or as an oxidizing agent to give $Ru(bpy)_3^+$. Therefore, the energy-transfer process may be coupled to either an oxidizing or a reducing agent. An oxidative coupling is shown in Scheme 7.4.

Scheme 7.4

Oxidative coupling

$$\{Ru(bpy)_3^{2+}\}^* + A \longrightarrow Ru(bpy)_3^{3+} + A^-$$

$$Ru(bpy)_3^{3+} + \text{Red. agent} \longrightarrow Ru(bpy)_3^{2+}$$

$$A^- + \text{Substrate} \xrightarrow{\text{Catalyst}} \text{Product} + A$$

Much of the interest in these systems has been concerned with the cleavage of water. For example, an oxidative system has been used[48] and studied in detail[49] in which A = $Rh(bpy)_3^{3+}$, triethanolamine is the reducing agent, H_2O is the substrate, Pt^0 is the catalyst, and H_2 is the product. A reductive scheme[50] uses A = ascorbate with $Co^{II}(bpy)_n$ as the oxidizing agent to form $Co^I(bpy)_n$. It is proposed that the latter reacts with H^+ to form a hydride that decomposes to H_2. A major problem in these applications is the destruction of $Ru(bpy)_3^{2+}$ by photoaquation. Balzani and co-workers[51] have recently prepared caged Ru(II) polypyridine complexes that are much more stable and have electronic properties similar to $Ru(bpy)_3^{2+}$.

7.7 ORGANOMETALLIC PHOTOCHEMISTRY

Photochemical conditions are widely used in synthetic organometallic studies, and there is a steadily increasing number of quantitative studies concerned with quantum yields and elucidation of the photochemical mechanism. Most of the systematic work has been done on metal carbonyls and their derivatives, and the following discussion will be limited to this class of compounds. A substantial barrier to mechanistic studies is the uncertainty about the assignment of the bands in the electronic spectrum of even the binary carbonyls. This problem has been discussed recently by Vanquickenborne and co-workers.[52] The electronic spectra are dominated by intense metal-to-ligand charge-transfer (MLCT) bands that are essentially transitions from the back-bonding $(d \rightarrow \pi^*CO)$ π orbitals to the corresponding π^* orbitals. In addition, there are weaker, essentially *d–d* transitions. The bands tend to overlap, with the latter being at somewhat longer wavelengths. The recent discussion indicates that *d–s* transitions may also be observed in the near-ultraviolet region for neutral and anionic $M(CO)_6$ compounds.

7.7.a   Metal Hexacarbonyls

Nasielski and Colas[53] studied the following reaction in benzene and cyclohexane with M = W:

$$M(CO)_6 + L \xrightarrow{\;h\nu\;} M(CO)_5L + CO \qquad (7.8)$$

They found that the quantum yield ($\Phi \approx 0.7$) is independent of wavelength between 254 and 366 nm and is independent of the concentration of L (= py or CH_3CN). The reaction is sensitized by $Ph_2C=O$ ($\Delta E^* = 289$ kJ mol^{-1} = 414 nm) but not by triphenylene ($\Delta E^* = 280$ kJ mol^{-1} = 427 nm). The reaction is not quenched by bibenzene ($\Delta E^* = 272$ kJ mol^{-1}) or naphthalene ($\Delta E^* = 255$ kJ mol^{-1}) and no phosphorescence was observed. The lack of quenching and phosphorescence indicates that the photoactive state is quite short-lived. The lack of concentration dependence is consistent with a dissociative mode of activation. The photoactive state was assigned to a triplet state involving metal *d* orbitals [see states for Co(III) in Figure 7.2] that shows a weak absorption at ~353 nm. This state presumably is reached by intersystem crossing when irradiation is at higher energies.

Flash photolysis of $Cr(CO)_6$ in hexane[54] produces the transient solvent complex $Cr(CO)_5 \cdot C_6H_{12}$, which was identified by infrared spectroscopy. This species reacts with CO and with H_2O with rate constants of 3.6×10^6 and 4.5×10^7 M^{-1} s^{-1} at 25°C, respectively, both with $\Delta H^* \approx 22$ kJ mol^{-1}. The reaction of $Cr(CO)_5OH_2$ with C_6H_{12} has k = 670 s^{-1} and $\Delta H^* \approx 75$ kJ mol^{-1}. The identification of the H_2O complex is of general interest because water is a potential contaminant in many studies. Similar studies[55] in aliphatic alcohols indicate initial formation of a CH-coordinated species that is rearranged to the OH-coordinated isomer with k $\approx 2 \times 10^{10}$ s^{-1}.

Other work[56,57] using picosecond laser spectroscopy has shown that these reactions proceed through a solvent intermediate, $M(CO)_5$(solvent), which forms in a few picoseconds after the laser pulse and then decays to products. Lee and Harris[58] have observed formation of the solvated species $Cr(CO)_5$(cyclohexane) with a lifetime $\tau = 17$ ps and the decay of the vibrationally excited $Cr(CO)_5$ with $\tau \approx 21$ ps (apparently at ambient temperature). These observations are at variance with those of Spears and co-workers,[59] who claim that the bare $Cr(CO)_5$ persists on the 100 ps time scale at 22°C. Hopkins and co-workers[60] have used resonance Raman detection to show that the 100 ps process is due to thermal relaxation of the excited vibrational state, probably of $Cr(CO)_5$(cyclohexane).

7.7.b Substituted Metal Carbonyls: $M(CO)_5L$

$M(CO)_5L$ systems were first studied by Wrighton et al.[61] and by Dahlgren and Zinck.[62] The quantum yields are in the range of 0.2 to 0.6. The photochemical reaction proceeds by two paths: L elimination (path A) and CO elimination (path B), as shown in Scheme 7.5.

Scheme 7.5

$$M(CO)_5L + Y \;\xrightarrow{h\nu}\; \begin{array}{l} \xrightarrow{\;A\;} M(CO)_5Y + L \\ \xrightarrow{\;B\;} \textit{cis}\text{-}M(CO)_4L(Y) + CO \end{array}$$

Wrighton et al. studied the system with $M = Mo$ and $L = Y = n\text{-PrNH}_2$, so that only path **B** is observed, and found that the quantum yield decreases with increasing irradiation wavelength ($\Phi_{366} = 0.24$; $\Phi_{405} = 0.20$; $\Phi_{436} = 0.057$). When $L = n\text{-PrNH}_2$ and $Y = 1$-pentene, path A is dominant and the quantum yield is less sensitive to wavelength, even increasing slightly at longer wavelength ($\Phi_{366} = 0.60$; $\Phi_{405} = 0.65$; $\Phi_{436} = 0.73$). The results were interpreted in terms of a ligand field model with photoactive triplet states, as shown in Figure 7.4. Longer wavelength radiation tends to populate the $d_{x^2-y^2}$ orbital and labilizes the cis-CO ligands.

Dahlgren and Zinck found that the importance of the two pathways depends on the nature of L. When L is a nitrogen-donor ligand (pyridine, piperidine, ammonia, acetonitrile), pathway A dominates. When L is a phosphine, the two pathways have about equal quantum yields, as shown by Eq. (7.9) and (7.10).

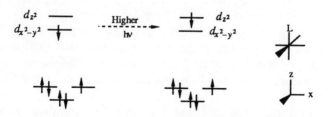

Figure 7.4. Ligand field excited states for $Mo(CO)_4LY$ complexes.

$$W(CO)_5N\text{—}R \xrightarrow[Y]{h\nu} \begin{array}{c} W(CO)_5Y \\ + N\text{—}R \\ \Phi \approx 0.5 \end{array} + \begin{array}{c} W(CO)_4(N\text{—}R)(Y) \\ + CO \\ \Phi < 0.01 \end{array} \qquad (7.9)$$

$$W(CO)_5P\text{—}R \xrightarrow[Y]{h\nu} \begin{array}{c} W(CO)_5Y \\ + P\text{—}R \\ \Phi \approx 0.3 \end{array} + \begin{array}{c} W(CO)_4(P\text{—}R)(Y) \\ + CO \\ \Phi \approx 0.3 \end{array} \qquad (7.10)$$

These observations were rationalized in terms of the M—C bond strengths as reflected by the CO stretching frequencies. Since the N-donors are not involved in back π bonding, the M—C bond is stronger and CO loss is not as favored as with the phosphines, which can compete with CO for π electrons on the metal, thereby weakening the M—C bond.

Darensbourg and Murphy[63] studied the photosubstitution of $Mo(CO)_5PPh_3$ with Y = PPH_3 or ^{13}CO (in THF at 313 and 366 nm). The results can be understood in terms of two five-coordinate intermediates, as shown in Scheme 7.6, where L = PPH_3.

Scheme 7.6

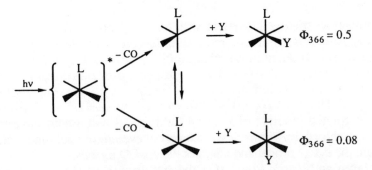

The interequilibration of the intermediates is indicated by the fact that *trans*-$Mo(CO)_4(PPh_3)_2$ photoisomerizes to the cis isomer with $\Phi_{366} = 0.3$.

Further work has been done by Lees and co-workers[64] on the Cr and Mo systems to try and elucidate the photoactive state. It was concluded for $Mo(CO)_5L$ (where L are various substituted pyridines), that the lowest energy excited state is either metal-to-ligand charge-transfer or ligand field in nature, depending on the nature of L. The charge-transfer state is readily deactivated and therefore less photoactive than the ligand field state. When the latter is lower in energy, it can be populated by intersystem crossing and gives larger and less wavelength-dependent quantum yields for replacement of L by PPh_3.

Recent pulsed-laser studies[65] with $W(CO)_5L$ (L = pyridine or piperidine) have essentially confirmed the earlier observations. This work also shows that a solvent complex is formed within 10 ps of the laser flash and this "intermediate" is the ultimate source of the substitution products. These authors also report that if 1-hexene is the entering ligand, then the first product has $W(CO)_5$ complexed to the "alkyl portion" of the hexene and this rearranges in about 10 ns to the η^2-hexene product. These observations may have some relevance to those of Stoutland and Bergman on the addition of ethylene to $IrCp^*(PMe_3)$ discussed in Section 5.4.

The volumes of activation for the photosubstitution of $W(CO)_5(py)$ and 4-substituted pyridines have been determined[66] using $P(OEt)_3$ as the entering group. This work also involved a study of the effect of pressure on the emission lifetimes, and the effect was found to be small compared to the effect of pressure on the quantum yield. The ΔV^* values are positive and somewhat dependent on the nature of the pyridine (py, 5.7 cm^3 mol^{-1}; 4-cyanopy, 6.3 cm^3 mol^{-1}; 4-acetylpy, 9.9 cm^3 mol^{-1}). The results are consistent with a dissociative mode of photoactivation.

7.7.c Manganese Pentacarbonyls

Faltynek and Wrighton[67] studied the photosubstitution of $Mn(CO)_5^-$ and $Mn(CO)_4PPh_3^-$ in the presence of PPh_3 or $P(OMe)_3$ and found quantum yields of ~ 0.3. In $Mn(CO)_4PPh_3^-$, only substitution of the PPh_3 is observed. These studies are complicated by the air sensitivity of the reactants and by secondary photolysis.

Oxidative addition to $Mn(CO)_5^-$ is also photoactivated. A competition study was done on substitution versus oxidative addition with the results shown in Scheme 7.7.

Scheme 7.7

$$(Ph)_4P^+ + Mn(CO)_5^- + PPh_3 \xrightarrow[\text{THF}]{h\nu} (Ph)Mn(CO)_4(PPh_3) + Mn(CO)_4PPh_3^-$$

M	M	M	Percent	Percent
0.01	0.01	0.01	35	38
0.01	0.01	0.06	31	32
0.01	0.01	0.22	28	82

The constancy of the oxidative addition yield suggests that this path may involve an ion pair. Unfortunately, the concentration of $P(Ph)_4^+$ was not varied to test this hypothesis.

Ford and co-workers[68] have studied the flash photolysis of $Mn(CO)_5CH_3$ at 308 nm in hydrocarbons and THF. They conclude that the transient $\{Mn(CO)_4CH_3\}$ reacts to give *cis*-(Solvent)$Mn(CO)_4CH_3$ and 5 x 10^3 faster with CO to give reactant. The complexes with the solvents C_6H_{12} and THF have rate constants with CO of 2 x 10^6 and 1.4 x 10^2 M^{-1} s^{-1}, respectively.

7.7.d   Mn₂(CO)₁₀ and Related Systems

The $Mn_2(CO)_{10}$ system has been of interest because it is a prototype for metal–metal bonded systems and because of the suggestion that •$Mn(CO)_5$ radicals may be involved in the thermal substitution reactions. The photochemistry of metal–metal bonded systems has been reviewed recently by Meyers and Caspar.[69] Bonding theory for $Mn_2(CO)_{10}$ suggests that irradiation can populate the σ^* orbital of the Mn—Mn bond and therefore might be expected to cleave that bond and produce the •$Mn(CO)_5$ radical. The $\sigma \rightarrow \sigma^*$ transition is observed at ~340 nm, and several studies[70,71] using 366-nm irradiation have found evidence for radical pathways for decomposition and substitution that are consistent with initial formation of •$Mn(CO)_5$. The final products in the presence of N-donor ligands are $Mn(CO)_5^-$ and $M(N)_6^{2+}$. A radical chain process has been proposed for the decomposition. The radical •$Mn(CO)_5$ reacts with CCl_4 via halogen atom abstraction $(k \approx 10^6 \ M^{-1} \ s^{-1})$ and this type of reaction is often used as a test for the presence of radicals.

Studies by Church et al.[72] have shown that the recombination of •$Mn(CO)_5$ occurs at a near diffusion-controlled rate $(k = 1 \times 10^9 \ M^{-1} \ s^{-1}$ in heptane). Infrared studies indicate that the radical has a square pyramidal structure. A species with a bridging CO has been identified from its infrared absorption at 1760 cm⁻¹ and is thought to have the following structure:

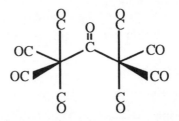

This species reacts with CO in C_7H_{16} to form $Mn_2(CO)_{10}$ with a rate constant of $2.7 \times 10^6 \ M^{-1} \ s^{-1}$. These observations have been confirmed by Seder et al.[73] in the gas phase, where the analogous rate constant is a surprisingly similar $2.4 \times 10^6 \ M^{-1} \ s^{-1}$ at 323 K and the radical recombination rate constant is $4.5 \times 10^{10} \ M^{-1} \ s^{-1}$ at 323 K. These results are of potential relevance to the thermal homolysis of the Mn—Mn bond.

Wrighton and co-workers[74] first reported that photolysis at 355 nm yields substitution of PPh₃, but later modified this[75] to about 30 percent CO dissociation and subsequent substitution. Recent work[76] on the wavelength dependence of the quantum yields indicates that CO dissociation becomes more important as the irradiation wavelength decreases from 355 to 255 nm, and this trend has been confirmed down to 193 nm in the gas phase by Seder et al. The situation may be summarized as in Scheme 7.8. The greater amount of CO dissociation at shorter wavelengths is consistent with irradiation into π^* orbitals of the M—CO bond.

Scheme 7.8

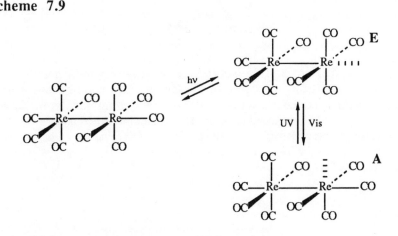

Turner and co-workers[77] have compared the photochemistry of $Mn_2(CO)_{10}$, $MnRe(CO)_{10}$, and $Re_2(CO)_{10}$ in Ar matrixes and liquid xenon. The former two are quite similar, but $Re_2(CO)_{10}$ is different in that it does not appear to form a bridged species. It is also known[78] that $Re_2(CO)_{10}$ gives a much greater percentage of CO dissociation at a given wavelength. The low-temperature studies in inert gases used infrared spectroscopy to reveal the intermediates shown in Scheme 7.9. The first intermediate (**E**), with an equatorial vacancy, is quite persistent in liquid xenon and returns to reactant over a period of several hours. The second intermediate (**A**), with an axial vacancy, is formed by irradiation of **E** by visible light and returns to **E** when ultraviolet light is used. Both intermediates react with N_2 to give the corresponding dinitrogen complex.

Scheme 7.9

7.7.e $M_3(CO)_{12}$ Systems: M = Fe, Ru, Os

These cluster compounds have been the subject of recent studies by Bentsen and Wrighton[79] and by Ford and co-workers[80] from which some general patterns have emerged. For wavelengths below about 350 nm, the dominant photoactivated process is CO loss and substitution by solvent or added nucleophiles. For wavelengths larger than 400 nm, the dominant process is photofragmentation in which the trinuclear species breaks into $M(CO)_5$, $M(CO)_4L$, $M_2(CO)_3L$, and so on. The reactions are unaffected by chlorinated organic solvents, and this is evidence against radical reaction pathways.

Bentsen and Wrighton proposed that short-wavelength irradiation of the Fe and Ru systems liberates an equatorial CO to give an intermediate that either captures solvent or added nucleophiles or rapidly rearranges to a more stable form with an axial vacancy. The latter rearranges to a bridged species identified by infrared bands in the region of 1830 cm^{-1}. The Os system is similar except that no bridged form appears to be present. The structures of these intermediates are shown in Scheme 7.10.

Scheme 7.10

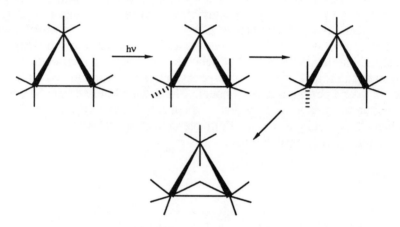

The study of Ford and co-workers on $Ru_3(CO)_{12}$ indicates that the intermediate $Ru_3(CO)_{11}$ produced by 308-nm irradiation in isooctane reacts with CO with a rate constant of 2×10^9 M^{-1} s^{-1} at ambient temperature. When THF is added to the solution, a THF adduct is formed and the kinetics of the reformation of $Ru_3(CO)_{12}$ are consistent with Scheme 7.11.

Scheme 7.11

$$Ru_3(CO)_{12} \xrightarrow{h\nu} Ru_3(CO)_{11} + CO$$

$$Ru_3(CO)_{11} + THF \underset{k_{-s}}{\overset{k_s}{\rightleftharpoons}} Ru_3(CO)_{11}(THF)$$

$$Ru_3(CO)_{11} + CO \xrightarrow{k_{CO}} Ru_3(CO)_{12}$$

From the dependence of the rate on [CO] and [THF] and the known value of k_{CO}, the authors obtain $k_{-s} = 2.1 \times 10^6$ s^{-1} and $k_s = 6 \times 10^9$ M^{-1} s^{-1}. The values of k_{CO} and k_s are near the diffusion-controlled limit and imply that isooctane is weakly solvating the intermediate $Ru_3(CO)_{11}$. The value of k_{CO} is much larger than that with $Mn_2(CO)_9$ in which a bridging CO is proposed to inhibit the reaction.

The long-wavelength process has been investigated by Bentsen and Wrighton,[79] by Poë and Sekhar,[81] and by Ford and co-workers.[82] The quantum yields are low (<0.02) and dependent on the nature of the added nucleophiles and the solvent. Product studies reveal that PPh$_3$ and P(OEt)$_3$ favor associative substitution to give M$_3$(CO)$_{11}$(PR$_3$), whereas octene and ethylene give fragmentation to M(CO)$_4$L. The results have been interpreted as indicating a common intermediate that is an unstable isomer of M$_3$(CO)$_{12}$, suggested to be the bridged isomer shown in Scheme 7.12. Ford has given a detailed analysis of the kinetics of this type of scheme in terms of the concentration dependence of the quantum yields for the Ru system.

Scheme 7.12

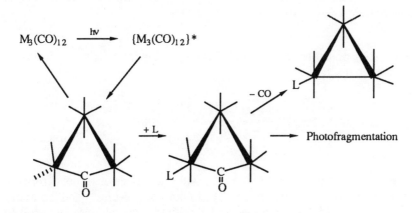

7.8 PHOTOCHEMICAL GENERATION OF REACTION INTERMEDIATES

An increasingly important application of photochemistry is in the production of proposed intermediates in thermal reactions. Flash photolysis and pulsed-laser techniques at low temperatures can produce these species and allow for their spectroscopic characterization and studies of their reactivity. Examples of this have been seen in the areas of C—H activation and the addition of alkenes to metal centers. The only caveat is that the species must be in the electronic and thermal ground state to be relevant to the thermal mechanism.

An example of this type of application is the photochemistry of the derivative of Wilkinson's catalyst, RhCl(CO)(PPh$_3$)$_2$, studied by Wink and Ford.[83] The flash photolysis was studied in benzene for wavelengths larger than 315 nm, and a transient species was observed with a much higher absorbance than the reactant in the 390- to 550-nm region. The transient species, which reacts with CO to produce the reactant, has been assigned as RhCl(PPh$_3$)$_2$. The reactivity of this transient with several species was studied and the results are summarized in Figure 7.5.

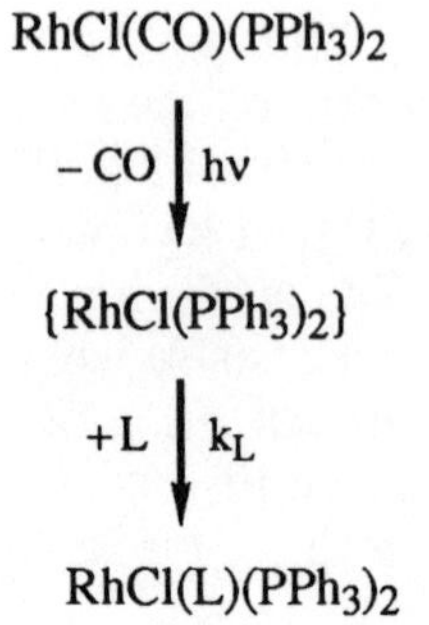

	L	k_L ($M^{-1} s^{-1}$)
$RhCl(CO)(PPh_3)_2$	CO	6.9×10^7
$-CO \downarrow h\nu$	C_2H_4	$>4 \times 10^7$
$\{RhCl(PPh_3)_2\}$	PPh_3	3.0×10^6
$+L \downarrow k_L$	H_2	1.0×10^5
	D_2	6.8×10^4
$RhCl(L)(PPh_3)_2$		

Figure 7.5. Substitution rate constants for the intermediate photochemically produced from a derivative of Wilkinson's catalyst.

It was also possible to determine the rate constant for the following dimerization reaction:

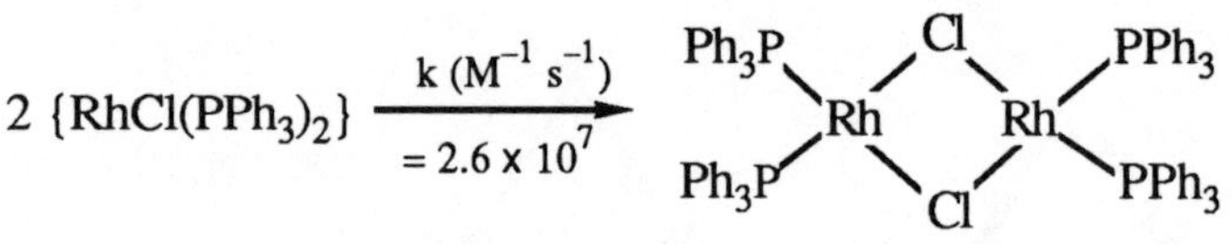

$$2 \{RhCl(PPh_3)_2\} \xrightarrow[= 2.6 \times 10^7]{k \ (M^{-1} s^{-1})} \begin{array}{c} Ph_3P \quad Cl \quad PPh_3 \\ Rh \quad Rh \\ Ph_3P \quad Cl \quad PPh_3 \end{array} \quad (7.11)$$

These results are in general agreement with the values or limits determined from conventional kinetic studies by Halpern and Wong[84] and Tolman et al.[85] It was found that the dimer and the dihydrogen adduct react with CO rather slowly ($k \approx 2 \ s^{-1}$) and by a unimolecular process. A remarkable feature of these results is the large rate constant for displacement of ethylene by CO, as shown in the following reaction:

$$RhCl(H_2C=CH_2)(PPh_3)_2 \xrightarrow[= 1 \times 10^8]{k \ (M^{-1} s^{-1})} RhCl(CO)(PPh_3)_2 \quad (7.12)$$
$$+ \ CO \qquad\qquad\qquad + \ H_2C=CH_2$$

Now, one has a quite complete picture of the reactivity of the species thought to be important in the operation of this catalytic system.

References

1. Balzani, V.; Carassiti, V. *Photochemistry of Coordination Compounds*; Academic Press: New York, 1970.
2. *Concepts of Inorganic Photochemistry*; Adamson, A. W.; Fleischauer, P., Eds.; Wiley: New York, 1975.
3. Geoffrey, G. L.; Wrighton, M. S. *Organometallic Photochemistry*; Academic Press: New York, 1979.

4. *Homogeneous and Heterogeneous Photocatalysis*; Pelizetti, E.; Serpone, N., Eds.; Reidel: Dordrecht, The Netherlands, 1986.

5. *Photochemistry and Photophysics of Coordination Compounds*; Yersin, H.; Vogler, A., Eds.; Springer-Verlag: Berlin, 1987.

6. Adamson, A. W.; Waltz, W. L.; Zinato, E.; Watts, D. W.; Fleischauer, P. D.; Lindholm, R. D. *Chem. Rev.* **1968**, *68*, 541.

7. Wrighton, M. S. *Top. Curr. Chem.* **1976**, *65*, 37.

8. Koerner von Gustof, E. A.; Linders, L. H. G.; Fischler, I.; Perutz, R. N. *Adv. Inorg. Chem. Radiochem.* **1976**, *19*, 65.

9. Crosby, G. A. *Acc. Chem. Res.* **1975**, *8*, 231.

10. Hollebone, B. R.; Langford, C. H.; Serpone, N. *Coord. Chem. Rev.* **1981**, *39*, 181.

11. Kirk, A. D. *Coord. Chem. Rev.* **1981**, *39*, 225.

12. Ford, P. C.; Wink, D.; DiBenedetto, J. *Prog. Inorg. Chem.* **1983**, *30*, 213.

13. Wegner, E. E.; Adamson, A. W. *J. Am. Chem Soc.* **1966**, *88*, 394.

14. Wilson, R. B.; Solomon, E. I. *J. Am. Chem. Soc.* **1980**, *102*, 4085.

15. Scandola, F.; Scandola, M. A.; Bartocci, C. *J. Am. Chem. Soc.* **1975**, *97*, 4757.

16. Nishazawa, M.; Ford, P. C. *Inorg. Chem.* **1981**, *20*, 294.

17. Wong, C. F. C.; Kirk, A. D. *Can. J. Chem.* **1976**, *54*, 3794.

18. Kirk, A. D.; Kneeland, D. M. *Inorg. Chem.* **1989**, *28*, 4274.

19. Endicott, J. F.; Ferraudi, G. J.; Barber, J. R. *J. Phys. Chem.* **1975**, *79*, 630; Endicott, J. F.; Ferraudi, G. J.; Barber, J. R. *J. Am. Chem. Soc.* **1975**, *97*, 219; Ferraudi, G. J.; Endicott, J. F.; Barber, J. R. *J. Am. Chem. Soc.* **1975**, *97*, 6406.

20. Weit, S. K.; Kutal, C. *Inorg. Chem.* **1990**, *29*, 1455.

21. Ford, P. C. *Inorg. Chem.* **1975**, *14*, 1440.

22. Bergkamp, M. A.; Watts, R. J.; Ford, P. C. *J. Am. Chem. Soc.* **1980**, *102*, 2627.

23. Weber, W.; van Eldik, R.; Kelm, H.; DiBenedetto, J; Ducommun, Y.; Offen, H.; Ford, P. C. *Inorg. Chem.* **1983**, *22*, 623.

24. Wieland, S.; DiBenedetto, J.; van Eldik, R.; Ford, P. C. *Inorg. Chem.* **1986**, *25*, 4893.

25. Kelly, T. L.; Endicott, J. F. *J. Am. Chem. Soc.* **1972**, *94*, 1797; *J. Phys Chem.* **1972**, *76*, 1937.

26. Zinato, E.; Lindholm, R. D.; Adamson, A. W. *J. Am. Chem. Soc.* **1969**, *91*, 1076.

27. Waltz, W. L.; Lillie, J.; Lee, S. H. *Inorg. Chem.* **1983**, *23*, 1768.

28. Wasgestian, H. F. *Z. Phys. Chem. N. F.* **1969**, *67*, 39.

29. Angermann, K.; van Eldik, R.; Kelm, H.; Wasgestian, F. *Inorg. Chim. Acta* **1981**, *49*, 247.

30. Riccieri, P.; Zinato, E. *Inorg. Chem.* **1990**, *29*, 5035.

31. Adamson, A. W. *J. Phys. Chem.* **1967**, *71*, 798.

32. Zinck, J. I. *J. Am. Chem. Soc.* **1974**, *96*, 4464; Ibid. **1972**, *94*, 8039.

33. Vanquickenborne, L. G.; Cuelemans, A. *Coord. Chem. Rev.* **1983**, *48*, 157.

34. Kirk, A. D.; Namasivayam, C.; Ward, T. *Inorg. Chem.* **1986**, *25*, 2225.

35. Mønsted, L.; Mønsted, O. *Coord. Chem. Rev.* **1989**, *94*, 109.

36. Mønsted, L.; Mønsted, O. *Acta Chem. Scand. A* **1986**, *40*, 637.

37. Kutal, C.; Adamson, A. W. *Inorg. Chem.* **1973**, *12*, 1990.

38. Kane-Maguire, N. A. P.; Wallace, K. C.; Miller, D. B. *Inorg. Chem.* **1985**, 24, 597.
39. Angermann, K.; van Eldik, R.; Kelm, H.; Wagestian, F. *Inorg. Chem.* **1981**, 20, 955.
40. Durham, B.; Caspar, J. V.; Nagle, J. K.; Meyer, T. J. *J. Am. Chem. Soc.* **1982**, 104, 4803.
41. Durham, B.; Walsh, J. L.; Carter, C. L.; Meyer, T. J. *Inorg. Chem.* **1980**, 19, 860.
42. Durham, B.; Wilson, S. R.; Hodgson, D. J.; Meyer, T. J. *J. Am. Chem. Soc.* **1980**, 102, 600.
43. Kalyanasundaram, K. *Coord. Chem. Rev.* **1982**, 46, 159.
44. Meyer, T. J. *Pure Appl. Chem.* **1986**, 58, 1193.
45. Juris, A.; Balzani, V.; Barigelleti, F.; Campagna, S.; Belser, P.; von Zelewsky, A. *Coord. Chem. Rev.* **1988**, 84, 85.
46. Krausz, E.; Ferguson, J. *Prog. Inorg. Chem.* **1989**, 37, 293.
47. Parris, T. P.; Brandt, W. W. *J. Am. Chem. Soc.* **1959**, 81, 5001.
48. Kirch, M.; Lehn, J.-M.; Sauvage, J.-P. *Helv. Chim. Acta* **1979**, 62, 1345; Lehn, J.-M.; Sauvage, J.-P. *Nouv. J. Chim.* **1981**, 5, 291.
49. Chan, S.-F.; Chou, M.; Creutz, C.; Matsubara, T.; Sutin, N. *J. Am. Chem. Soc.* **1981**, 103, 369.
50. Krishnan, C. K.; Sutin, N. *J. Am. Chem. Soc.* **1981**, 103, 2140.
51. Barigelletti, F.; De Cola, L.; Balzani, V.; Belser, P.; von Zelewsky, A.; Vögtle, F.; Ebmeyer, F.; Grammenudi, S. *J. Am. Chem. Soc.* **1989**, 111, 4662.
52. Pierloot, K.; Verhulst, J.; Verbeke, P.; Vanquickenborne, L. G. *Inorg. Chem.* **1989**, 28, 3059.
53. Nasielski, J.; Colas, A. *Inorg. Chem.* **1978**, 17, 237.
54. Church, S. P.; Grevels, F.-H.; Hermann, H.; Schaffner, K. *Inorg. Chem.* **1985**, 24, 418; Ibid. **1984**, 23, 3830.
55. Xie, X.; Simon, J. D. *J. Am. Chem. Soc.* **1990**, 112, 1130.
56. Langford, C. H.; Moralejo, C.; Sharma, D. K. *Inorg. Chim. Acta* **1987**, 126, L11.
57. Simon, J. D.; Xie, X. *J. Phys. Chem.* **1987**, 91, 5538; Ibid. **1989**, 93, 291.
58. Lee, M.; Harris, C. B. *J. Am. Chem. Soc.* **1989**, 111, 8963.
59. Wang, L.; Zhu, X.; Spears, K. G. *J. Phys. Chem.* **1989**, 93, 2; *J. Am. Chem. Soc.* **1988**, 110, 8695.
60. Yu, S.-C.; Xu, X.; Lingle, R., Jr.; Hopkins, J. B. *J. Am. Chem. Soc.* **1990**, 112, 3668.
61. Wrighton, M. S.; Morse, D. L.; Gray, H. B.; Ottesen, D. K. *J. Am. Chem. Soc.* **1976**, 98, 1111.
62. Dahlgren, R. M.; Zinck, J. I. *Inorg. Chem.* **1977**, 16, 3154.
63. Darensbourg, D. J.; Murphy, M. A. *J. Am. Chem. Soc.* **1978**, 100, 463.
64. Lees, A. J. *J. Am. Chem. Soc.* **1982**, 104, 2038; Kolodziej, R. M.; Lees, A. J. *Organometallics* **1986**, 5, 450.
65. Moralejo, C.; Langford, C. H.; Sharma, D. K. *Inorg. Chem.* **1989**, 28, 2205.
66. Wieland, S.; van Eldik, R.; Crane, D. R.; Ford, P. C. *Inorg. Chem.* **1989**, 28, 3663.
67. Faltynek, R. A.; Wrighton, M. S. *J. Am. Chem. Soc.* **1978**, 100, 2701.

68. Belt, S. T.; Ryba, D. W.; Ford, P. C. *Inorg. Chem.* **1990**, *29*, 3633.
69. Meyers, T. J.; Caspar, J. V. *Chem. Rev.* **1985**, *85*, 187.
70. Stiegman, A. E.; Tyler, D. R. *Inorg. Chem.* **1984**, *23*, 527.
71. McCullen, S. B.; Brown, T. L. *Inorg. Chem.* **1981**, *20*, 3528.
72. Church, S. P.; Hermann, H.; Grevels, F.-W.; Schaffner, K. J. *J. Chem. Soc., Chem. Commun.* **1984**, 785; Church, S. P.; Poliakoff, M.; Timmey, J. A.; Turner, J. J. *J. Am. Chem. Soc.* **1981**, *103*, 7515.
73. Seder, T. A.; Church, S. P.; Weitz, E. *J. Am. Chem. Soc.* **1986**, *108*, 7518.
74. Wrighton, M. S.; Ginley, D. S. *J. Am. Chem. Soc.* **1975**, *97*, 2065.
75. Hepp, A. F.; Wrighton, M. S. *J. Am. Chem. Soc.* **1983**, *105*, 5934.
76. Kobayashi, T.; Yasufuku, K.; Iwai, J.; Yesaka, H.; Noda, H.; Ohtani, H. *Coord. Chem. Rev.* **1985**, *64*, 1.
77. Firth, S.; Klotzbücher, W. E.; Poliakoff, M.; Turner, J. J. *Inorg. Chem.* **1987**, *26*, 3370.
78. Kobayashi, T.; Ohtani, H.; Noda, H.; Teratani, S.; Yamazaki, H.; Yasufuku, K. *Organometallics* **1986**, *5*, 110.
79. Bentsen, J. G.; Wrighton, M. S. *J. Am. Chem. Soc.* **1987**, *109*, 4518, 4530.
80. DiBenedetto, J. A.; Ryba, D. W.; Ford, P. C. *Inorg. Chem.* **1989**, *28*, 3503.
81. Poë, A. J.; Sekhar, C. V. *J. Am. Chem. Soc.* **1986**, *108*, 3673.
82. Desrosiers, M. F.; Wink, D. A.; Trautman, R.; Friedman, A. E.; Ford, P. C. *J. Am. Chem. Soc.* **1986**, *108*, 1917.
83. Wink, D. A.; Ford, P. C. *J. Am. Chem. Soc.* **1987**, *109*, 436.
84. Halpern, J.; Wong, S. W. *J. Chem. Soc., Chem. Commun.* **1973**, 629.
85. Tolman, C. A.; Meakin, P. Z.; Lindner, D. L.; Jesson, J. P. *J. Am. Chem. Soc.* **1974**, *96*, 2762.

8

Bioinorganic Systems

The field of bioinorganic chemistry has grown tremendously in the past 20 years. Much of the work is concerned with establishing the coordination site, ligand geometry, and metal oxidation state in biologically active systems. The field also extends to the preparation and characterization of simpler model complexes that mimic the spectroscopic properties and perhaps some of the reactivity of the biological system. Much of this characterization work must precede meaningful mechanistic studies. Williams[1] has provided an interesting overview of metal ions in biology from an inorganic perspective. There are several review series[2] and specialized journals[3] devoted to the subject.

The field is so large and the systems are so individualistic that it is necessary, for the purposes of a text such as this, to choose a few sample systems as illustrative of the mechanistic achievements and problems.

8.1 VITAMIN B_{12}

In biological systems, Vitamin B_{12} consists of an enzyme–coenzyme complex which is called the *holoenzyme*. The enzyme is a peptide, whose composition depends on the biological source and typically has a molar mass in the range of 150,000 to 500,000 daltons. The coenzyme is a cobalt complex that can be isolated by denaturation of the peptide.

Coenzyme B_{12}, often called adenosylcobalamin or $AdoB_{12}$, is a cobalt(III) complex whose structure is shown in Figure 8.1. In the derivative that was first characterized, the 5'-deoxyadenosyl was replaced by cyanide as a result of the isolation procedure. This Co(III) compound is commonly called vitamin B_{12} or cyanocobalamin. If the 5'-deoxyadenosyl ligand is replaced by water, the compound is called aquocobalamin or B_{12a}. Methylcobalamin has a methyl group in place of the 5'-deoxyadenosyl. The cobalt(III) in $AdoB_{12}$ can be reduced to the cobalt(II) derivative, called cob(II)alamin or B_{12r}, and further reduction gives the cobalt(I) species, B_{12s}. These forms are all low-spin and can be distinguished by their electronic spectra, and B_{12r} is epr active. In coenzyme B_{12}, the benzimidazole may be displaced by solvent and protonated (pK_a 5 – 6) to give the base off form. The benzimidazole function can be removed by hydrolysis at the phosphate–CH(CH₃) linkage to give a series of complexes known as cobinamides. The chemistry and biochemistry of coenzyme B_{12} are the subject of a recent compilation.[4]

The structure[5] and coordination of coenzyme B_{12} are unusual in several aspects. It is an organometallic cobalt(III) complex with a cobalt–carbon bond to the 5'-carbon of 5'-deoxyadenosine. Four coordination positions in a plane are occupied by nitrogens from a corrin ring. This ring differs from the more common porphyrin in that two of the pyrrole rings are directly bonded, rather than having an intervening -CH group. The corrin ring is somewhat flexible and puckered to varying degrees, as discussed in detail by Pett et al.[6] If the 5'-deoxyadenosyl is replaced by a methyl group,[7] then the Co—C bond shortens from 2.05 to 1.99 Å and the Co—N(benzimidazole)

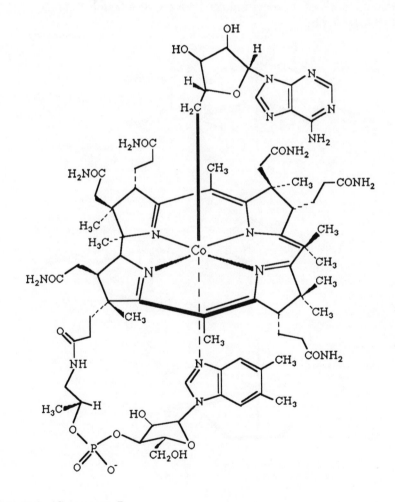

Figure 8.1. Coenzyme B_{12}.

also shortens from 2.24 to 2.19 Å. The crystal structure[8] indicates that the cobalt(II) derivative is five-coordinate, with the cobalt 0.12 Å below the plane of the corrin nitrogens toward the benzimidazole. The bond to the benzimidazole nitrogen is shorter in the cobalt(II) derivative (2.13 Å) than in coenzyme B_{12} (2.24 Å). An EXAFS study[9] has suggested that the benzimidazole-N—Co(II) bond is even shorter (1.99Å).

There are a number of simple model complexes of cobalt that mimic various aspects of the chemistry of coenzyme B_{12}, such as $Co(DH)_2$,[10] $Co(C_2(DO)(DOH))_{pn}$,[11] and Co(salen),[12] which are shown in Figure 8.2. These complexes have derivatives with cobalt in the (III), (II), and (I) oxidation states analogous to coenzyme B_{12}. They form a wide range of organometallic cobalt(III) species in which both the organic group and the sixth ligand can be varied. The models have been particularly useful for the characterization of the electronic and epr spectral features of coenzyme B_{12} and its derivatives.

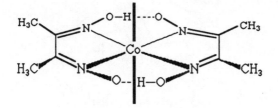

$Co(DH)_2$

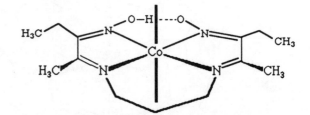

$Co(C_2(DO)(DOH))_{pn}$

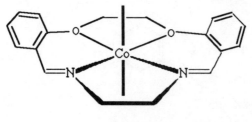

Co(salen)

Figure 8.2. Some coenzyme B_{12} model systems.

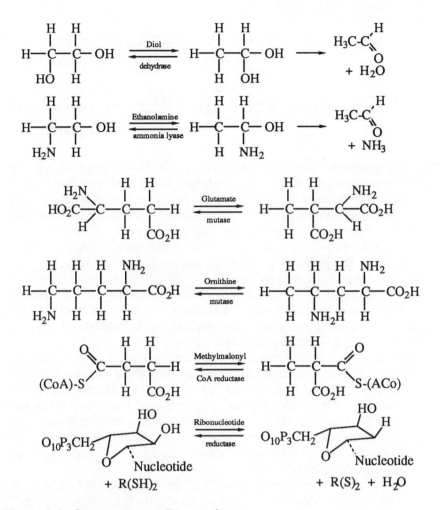

Figure 8.3. Some coenzyme B_{12} reactions.

Some enzymic reactions that occur with coenzyme B_{12} are given in Figure 8.3. The first five examples have as a common feature the interchange of H and another substituent X on an adjacent carbon atom (X = OH, NH_2, $CH(NH_2)CO_2H$, and (CoA)—S representing coenzyme A). Tritium and deuterium labeling experiments have established that a hydrogen of the substrate is transferred to the 5'-carbon of 5'-deoxyadenosine during the reactions of diol dehydrase.[13] This system gives inversion at the C-2 carbon when 1,2-propanediol is the substrate, and glutamate mutase also gives inversion. The reactions of methylmalonyl–CoA mutase and ribonucleotide reductase proceed with retention. The labeling experiments indicate that the initial reaction with a substrate RH may occur by a radical mechanism, as in Eq. (8.1), or by an ionic mechanism, as in Eq. (8.2):

$$Co^{III}\text{—}CH_2Ado \rightleftharpoons \underset{+ \bullet CH_2Ado}{Co^{II}} \overset{RH}{\rightleftharpoons} \underset{+ CH_3Ado}{Co^{II} + R\bullet} \qquad (8.1)$$

$$Co^{III}\text{—}CH_2Ado \rightleftharpoons \underset{+ {}^-CH_2Ado}{Co^{I}} \overset{RH}{\rightleftharpoons} \underset{+ CH_3Ado}{Co^{I} + R^-} \qquad (8.2)$$

The radical mechanism involves homolysis of the Co—C bond and the ionic mechanism involves heterolysis of this bond. Since coenzyme B_{12} has both Co^{II} and Co^{I} derivatives, either of these mechanisms is chemically reasonable. In addition, the substrate radical or anion might combine with the cobalt to give an organocobalt complex as the product of the second step in both (8.1) and (8.2). Several lines of evidence favor the radical mechanism. The presence of organic radicals and cobalt(II) as B_{12r} has been established by epr when substrate is added to the ethanol ammonia lyase,[14] diol dehydrase,[15] and ribonucleotide reductase systems,[16] and the rate of formation of these radicals was shown to be consistent with their being on the reaction pathway. However, with methylmalonyl coenzyme A mutase, no radicals have been detected.[17]

Although these reactions may not be proceeding by a common mechanism, much of the recent work has concentrated on the radical pathway because it seems to have solid support for some systems, as noted earlier. There are two chemical problems with the radical mechanism. The first is the thermal stability of the Co—C bond compared to the high reactivity of the enzymic reactions. The second is the very limited precedents for the type of radical rearrangements that are required. Various aspects and approaches to these problems are summarized in recent reviews.[18,19]

There have been a number of studies of the Co—C homolysis rate and bond energy in model compounds and more recently in coenzyme B_{12} itself. Halpern et al.[20] studied the temperature dependence of the equilibrium constant for reaction (8.3) and determined that ΔH° is 22.1 kcal mol^{-1}.

$$(py)(DH)_2Co\text{—}CH(CH_3)(C_6H_5) \overset{Toluene}{\rightleftharpoons} (py)(DH)_2Co^{II} \qquad (8.3)$$
$$+ C_6H_5CH=CH_2 + 1/2\ H_2$$

They combined this value with the ΔH° of -2.2 kcal mol^{-1} for reaction (8.4) to obtain a ΔH° of 19.9 kcal mol^{-1} for reaction (8.5). The latter is defined as the Co—C bond energy of the reactant in Eq. (8.3).

$$C_6H_5CH=CH_2 + 1/2\ H_2 \rightleftharpoons C_6H_5\overset{\bullet}{C}H(CH_3) \qquad (8.4)$$

The rate of reaction (8.3) in the absence of H_2 is first-order in the reactant, and it was assumed that the rate-controlling step is reaction (8.5):

$$(py)(DH)_2Co—CH(CH_3)(C_6H_5) \rightleftharpoons (py)(DH)_2Co^{II} \qquad (8.5)$$
$$+ C_6H_5\overset{\bullet}{C}H(CH_3)$$

with a forward rate constant $k_f = 7.8 \times 10^{-4}$ s^{-1} (25°C), $\Delta H_f^* = 21.2$ kcal mol^{-1}, and $\Delta S_f^* = -1.4$ cal mol^{-1} K^{-1}. Other work[21] indicates that the reverse of reaction (8.5) will be near the diffusion-controlled limit and therefore should have a low value of $\Delta H_r^* \sim 2$ kcal mol^{-1}. These results predict the $\Delta H°$ for reaction (8.5) as $21.2 - 2 = 19.2$ kcal mol^{-1}, in agreement with the value from the equilibrium measurements. This provides the basis for the determinations of a number of Co—C bond energies by measurement of the homolysis kinetics, especially if a radical scavenger can be added to drag the reaction to completion.

Variations in the values of $\Delta H°$ for reactions such as (8.5) with different axial bases and alkyl groups may be due to other bond energy differences in reactants and products and not just to the Co—C bond energy changes. Measurements[22] of the Co—C stretching frequency (~ 500 cm^{-1}) in L(DH)$_2$Co—CH$_3$ systems have been interpreted to indicate that apparent bond energy variations with L are more reflective of changes in the stability of the Co(II) product than of the Co(III) reactant. The applications and complications of the kinetic method for these bond energy determinations are the subject of several recent discussions.[23,24] In the absence of trapping agents, the radical from thermolysis or photolysis may react with the planar N$_4$ ligand system to give a stable product as observed[25] with the benzyl derivative of the Co(C$_2$(DO)(DOH))$_{pn}$ model, for example.

The temperature dependence of the thermal decomposition of coenzyme B$_{12}$ was studied first in ethylene glycol by Finke and Hay[26], then in water by Halpern et al.,[27] and most recently again in water by Hay and Finke.[28] The latter authors have criticized the earlier aqueous study for failure to separate all the reaction pathways, and in particular the heterolysis reaction, which is significant for the base off form of the coenzyme, and is found to contribute between 77 percent (85°C) and 45 percent (110°C) of the product at the pH = 4.3 of the earlier study. From results between 85°C and 110°C at pH 7, after correction for heterolysis and the base off form, Finke and Hay find the kinetic parameters for the homolysis to be $\Delta H_h^* = 33 \pm 2$ kcal mol^{-1} and $\Delta S_h^* = 11 \pm 3$ cal mol^{-1} K^{-1}, which give $k_h = 1 \times 10^{-9}$ s^{-1} (25°C). The Co—C bond dissociation energy is then estimated to be ~ 30 kcal mol^{-1} after correction for the activation energy for the reverse reaction. Martin and Finke[29] have used a similar method to estimate a Co—C bond dissociation energy of 37 kcal mol^{-1} for methylcobalamin.

The small value of k_h for the coenzyme is noteworthy in comparison to values of $\sim 2 \times 10^2$ s^{-1} estimated for the rate-determining step in the diol dehydrase and ethanolamine ammonia lyase systems. This shows one of the problems in using the radical mechanism. This difficulty usually is explained by assuming that the enzyme somehow distorts the coenzyme so that the homolysis is about 15 kcal mol^{-1} more favorable in the holoenzyme. This distortion may destabilize the reactant and/or stabilize the transition state in order to hasten homolysis in the holoenzyme.

The problem of appropriate precedents for the organic rearrangements required by the radical mechanism has been addressed and discussed by Wollowitz and Halpern.[30] They reacted (n-Bu)$_3$SnH with various bromide derivatives to generate the radicals and obtained some rearranged product, as shown for typical conditions in Scheme 8.1. The product ratios were determined as a function of (n-Bu)$_3$SnH concentration in order to determine the relative rates of rearrangement and H• abstraction from (n-Bu)$_3$SnH by the parent radical.

Scheme 8.1

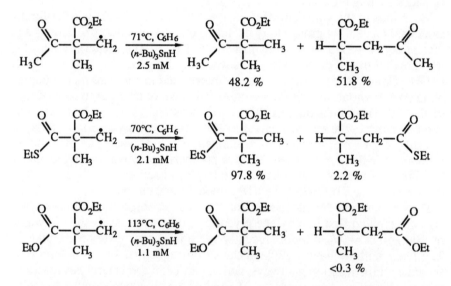

It is clear from the product amounts in Scheme 8.1 that the migratory aptitudes of the groups are O=CCH$_3$ > O=CSEt > O=COEt. The rearrangement is envisaged as proceeding through the following cyclopropyloxy intermediate or transition state:

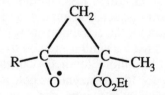

Rearrangement with the O=CSEt group seems relevant to the methylmalonyl–CoA reductase system, but there are still no precedents for the OH and NH_2 migration. In addition, Wollowitz and Halpern observed rearrangements in the corresponding anions that would result from heterolysis [Eq. (8.2)].

Choi and Dowd[31] have reported substantial amounts of rearranged products for reactions (8.6) to (8.8):

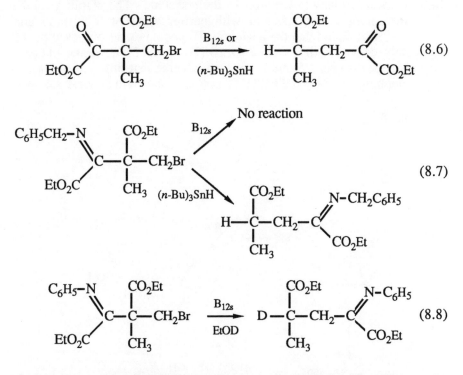

These reactions are expected to proceed by nucleophilic attack of B_{12s} on the halide to give the organo-B_{12} derivative that would undergo homolysis to products. However, the reactivity patterns with B_{12s} and the formation of the deuterated product in EtOD are more easily rationalized if the reactions are proceeding through an anionic organic intermediate, which might be produced by reduction of the organic radical by excess B_{12s} or other reducing agents used to prepare the B_{12s}. Murakami and et al.[32] have noted that anionic rearrangements are more facile than those involving radicals. They observed that addition of cyanide to organo-B_{12} derivatives yields the organic anion under photolysis conditions because the NC—Co^{II} homolysis product reduces the radical to the anion, which then is rearranged.

Dowd et al.[33] found that no rearranged product was obtained for the B_{12} derivative Co—$CH_2CH(CO_2H)CH(CO_2^-)(NH_3^+)$ under thermal or photochemical decomposition. However, Murakami et al.[34] observed substantial rearrangement in the organic products with the same and several related species when photolysis is done while the complex is bound to an anionic surfacant (octopus azaparacyclophane). They proposed that the surfactant is acting in a way analogous to the enzyme in the biological system. These results indicate that studies on the coenzyme alone may not be appropriate models for the enzyme.

Finke et al.[35] found that, with the model $Co(C_2(DO)(DOH))_{pn}$ in Figure 8.2, the Co—$CH_2CH(=O)$ complex was too stable to be an intermediate in the diol to acetaldehyde reaction in their system. The stability of the aldehyde complex is consistent with earlier work of Silverman and Dolphin,[36] who found that the analogous B_{12} complex decomposed by an acid-catalyzed path with $k_{obsd} = 2.1 \times 10^3[H^+]$ s^{-1} at 25°C. Finke and co-workers appear to have found a synthetic scheme that proceeds at least in part through the organometallic diol complex, as shown in Scheme 8.2.

Scheme 8.2

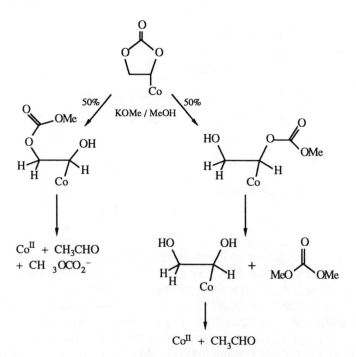

The intermediates in Scheme 8.2 apparently are too unstable to be isolated and have been identified through the nature and stoichiometry of their decomposition products ($Co(II) : CH_3CHO : CH_3OCO_2^- = 2 : 2 : 1$) and radical trapping experiments. The decomposition is too fast to be consistent with the formation of the relatively stable aldehyde complex as an intermediate. The observations have been interpreted by assuming that the radical formed after homolysis escapes from the solvent cage and undergoes a normal radical decomposition as shown in reaction (8.9), without the intervention of a cobalt complex:

$$\underset{\underset{H_2C-CH}{|\quad\quad|}}{\overset{HO\quad OH}{}} \xrightarrow{-H_2O} H_2\overset{\bullet}{C}-C\overset{O}{\underset{H}{\diagup}} \xrightarrow{\cdot H} H_3C-C\overset{O}{\underset{H}{\diagup}} \quad (8.9)$$

However, this mode of decomposition does not involve the 1,1-diol intermediate shown to be required by the elegant stereochemical and labeling studies of Arigoni and co-workers[37] on diol dehydrase. Finke et al.[19] have proposed a bound radical mechanism for the enzyme in which formation of the 1,1-diol assists in binding the radical to the enzyme by hydrogen bonding. The bound radical mechanism involves rehydration of the radical intermediate in reaction (8.9) to give the 1,1-diol.

A probable function of the enzyme is indicated by flash photolysis studies. With nanosecond time scale detection, Chen and Chance[38] found that coenzyme B_{12} undergoes homolysis to give a geminate pair of cobalt(II) and adenosyl radical with a quantum yield of 0.23 for the base on form. The geminate pair undergoes very rapid recombination with a rate constant larger than 10^8 s^{-1}. To counteract this recombination in the natural system, the enzyme may serve to move the adenosyl radical away from the cobalt(II).

8.2 A ZINC(II) ENZYME: CARBONIC ANHYDRASE

Zinc is the second most abundant metal, after iron, in humans. There are a number of enzyme systems in which zinc(II) is bound to a peptide (apoenzyme) and is at the active site in the enzyme. Examples of these enzymes are carbonic anhydrase, carboxypeptidase A and B, alkaline phosphatase, alcohol dehydrogenase, and RNA polymerase. In most cases, these systems bring about the making or breaking of covalent bonds in an organic substrate, but they may also use an oxidizing or reducing coenzyme to produce a redox change of the substrate. The area has been the subject of a recent compilation.[39] Studies on carbonic anhydrase are representative of the methods used to elucidate a considerable range of enzyme mechanisms.

Kinetic studies are almost inevitably an important part of the information used to characterize these systems. Most of the kinetic work has been done by biochemists, and they have evolved their own terminology for describing the results. The following section provides an outline of some of these terms and the general methodology.

8.2.a Introduction to Terms and Methods of Enzyme Kinetics

The simplest picture of an enzymic reaction is given by Eq. (8.10):

$$E + S \underset{k_2}{\overset{k_1}{\rightleftharpoons}} ES \underset{k_4}{\overset{k_3}{\rightleftharpoons}} E + P \quad (8.10)$$

where E is the enzyme, S is the substrate, ES is the enzyme–substrate complex, and P is the product. These systems are often studied under steady-state conditions in which the enzyme (usually $<10^{-6}$ M) and substrate (usually $>10^{-4}$ M) are mixed in a buffered solution and the initial rate of loss of S or formation of P is determined. The formation of ES is assumed to be a rapid pre-equilibrium, and only the initial rate is studied so that P does not accumulate and the k_4 step is ignored. This system was analyzed in Section 2.2 and the rate is given by the *Michaelis–Menton equation*:

$$\frac{d\,[P]}{dt} = v_i = \frac{k_3\,[E]_T\,[S]}{[S] + \dfrac{k_2}{k_1}} = \frac{k_3\,[E]_T\,[S]}{[S] + K_m} \tag{8.11}$$

where v_i is the initial rate, $[E]_T = [E] + [ES]$ is the total enzyme concentration and K_m is the *Michaelis constant*. Since $K_m = [E][S]/[ES]$, it would be called a dissociation constant by inorganic chemists and smaller values of K_m mean stronger enzyme to substrate binding. This rate law gives what inorganic kineticists call saturation behavior, and the rate reaches a maximum when $[S] \gg K_m$, so that $V_{max} = k_3[E]_T$. The ratio $V_{max}/[E]_T = k_3$ is called the *turnover number*. Since the total enzyme concentration may not be precisely known, it is useful to substitute V_{max} into Eq. (8.11) to obtain the more conventional form of the Michaelis–Menton equation. This equation can be rearranged by taking the reciprocal of both sides to obtain

$$\frac{1}{v_i} = \frac{K_m}{V_{max}\,[S]} + \frac{1}{V_{max}} \tag{8.12}$$

which predicts that a plot of v_i^{-1} versus $[S]^{-1}$, called a *Lineweaver–Burke plot*, will be linear with an intercept of $-K_m^{-1}$ when $v_i^{-1} = 0$.

If a steady-state approximation for ES is used (see Section 2.1) instead of the rapid-equilibrium assumption, then one obtains Eq. (8.13), known as the *Briggs–Haldane equation*:

$$\frac{d\,[P]}{dt} = v_i = \frac{k_3\,[E]_T\,[S]}{[S] + \dfrac{k_2 + k_3}{k_1}} \tag{8.13}$$

This equation has the same mathematical form as Eq. (8.11), except that $K_m = (k_2 + k_3)/k_1$. The K_m will be the same as defined previously only if $k_2 \gg k_3$; otherwise, K_m is always larger than k_2/k_1.

The variation of the rate of the enzyme-catalyzed reaction with pH can provide information about the ionization of functional groups important for the reaction. For example, the system might be described by Scheme 8.3, which has been simplified from more general possibilities in order to give a more tractable result.

Scheme 8.3

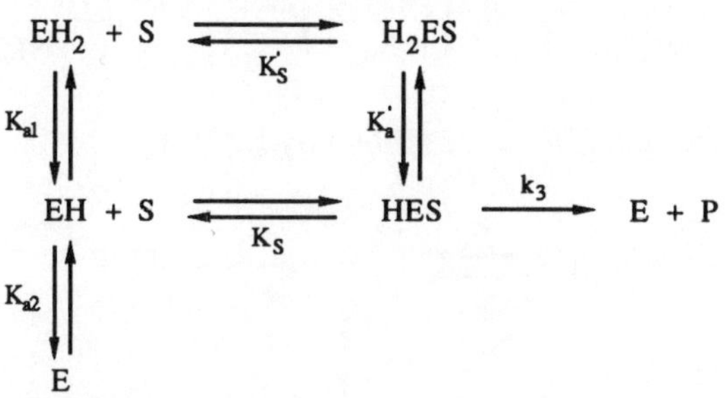

$$
\begin{array}{ccc}
EH_2 + S & \xrightleftharpoons[K_s']{} & H_2ES \\
K_{a1} \Updownarrow & & K_a' \Updownarrow \\
EH + S & \xrightleftharpoons[K_s]{} & HES \xrightarrow{k_3} E + P \\
K_{a2} \Updownarrow & & \\
E & &
\end{array}
$$

If it is assumed that the substrate binding and proton equilibria are rapidly maintained, then the initial rate can be put in the standard Michaelis–Menton form:

$$\frac{v_i}{[E]_T} = \frac{k_{cat}[S]}{[S] + K_m} \tag{8.14}$$

where

$$k_{cat} = \frac{k_3}{\left(1 + \dfrac{[H^+]}{K_a'}\right)} \quad \text{and} \quad K_m = K_s \frac{\left(1 + \dfrac{[H^+]}{K_{a1}} + \dfrac{K_{a2}}{[H^+]}\right)}{\left(1 + \dfrac{[H^+]}{K_a'}\right)}$$

If the conditions are such that $[S] \gg K_m$ (called swamping conditions), then the variation of k_{cat} with pH can be used to determine K_a'. If $K_m \gg [S]$, then k_{cat}/K_m is independent of K_a' and K_{a1} and K_{a2} can be determined.

Many enzyme systems are affected by inhibitors. The structure of the inhibitors may be helpful in defining the size and binding requirements of the active site, but they may bind elsewhere and cause distortions of the peptide that affect the active site. The kinetic effect of inhibitors can be described by three possibilities, and mixtures of these possibilities may be observed.

A system with a *competitive inhibitor* (I) is described by Eq. (8.15):

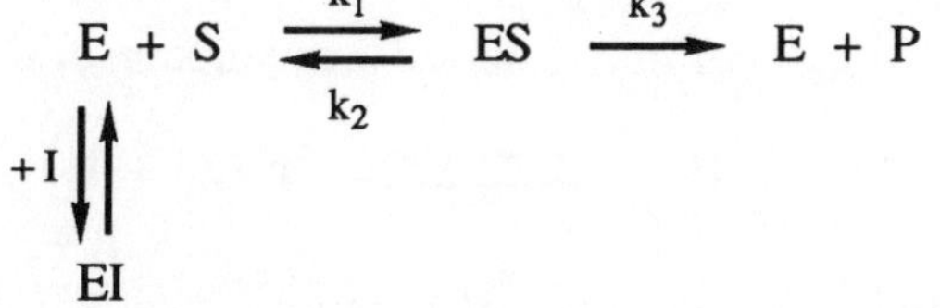

$$
\begin{array}{ccc}
E + S & \underset{k_2}{\overset{k_1}{\rightleftharpoons}} ES \xrightarrow{k_3} E + P & \tag{8.15} \\
{}_{+I} \Updownarrow & & \\
EI & &
\end{array}
$$

If a steady state is assumed for ES and $K_I = [E][I]/[EI]$ is a rapidly maintained equilibrium, then

$$v_i = \frac{k_3 [E]_T [S]}{[S] + \dfrac{(k_2 + k_3)}{k_1}\left(\dfrac{K_I + [I]}{K_I}\right)} = \frac{k_3 [E]_T [S]}{[S] + K_m\left(\dfrac{K_I + [I]}{K_I}\right)} \qquad (8.16)$$

An *uncompetitve inhibitor* is described by Eq (8.17):

$$E + S \underset{k_2}{\overset{k_1}{\rightleftharpoons}} ES \overset{k_3}{\longrightarrow} E + P \qquad (8.17)$$

$$+I \updownarrow$$

$$ESI$$

If the same kinetic assumptions are used, then

$$v_i = \frac{k_3 [E]_T [S]\left(\dfrac{K_I}{K_I + [I]}\right)}{[S] + \dfrac{(k_2 + k_3)}{k_1}\left(\dfrac{K_I}{K_I + [I]}\right)} = \frac{k_3 [E]_T [S]\left(\dfrac{K_I}{K_I + [I]}\right)}{[S] + K_m\left(\dfrac{K_I}{K_I + [I]}\right)} \qquad (8.18)$$

A *noncompetitive inhibitor* binds to both E and ES. If it is assumed that K_I is the same for both species, then

$$v_i = \frac{k_3 [E]_T [S]\left(\dfrac{K_I}{K_I + [I]}\right)}{[S] + \dfrac{(k_2 + k_3)}{k_1}} \qquad (8.19)$$

The various types of inhibition can be distinguished by studies of v_i for different S and I concentrations. Then Lineweaver–Burke plots of v_i^{-1} versus $[S]^{-1}$ at constant [I] or *Dixon plots* of v_i^{-1} versus [I] at constant [S] can be used to determine K_I and the type of inhibition.

8.2.b   Mechanism of Carbonic Anhydrase Action

The carbonic anhydrase enzymes catalyze the hydration of CO_2 shown by

$$CO_2 + H_2O \rightleftharpoons HCO_3^- + H^+ \qquad (8.20)$$

They also catalyze the hydrolysis of organic esters and the hydration of aldehydes, but with such low efficiency that this does not seem to be a significant biological function. Various aspects have been reviewed recently by Silverman and Lindskog.[40]

In mammals, there are three distinct isozymes of carbonic anhydrase, designated as CA I, CA II, and CA III. Their maximum turnover numbers at 25°C are 2×10^5, 1×10^6, and 3×10^3 s^{-1}, respectively, and CA II is one of the most efficient of all known enzymes. Although they have different reactivities and amino acid sequences, the structures of CA I and II are known to be homologous. The active site is typically a conical cleft about 15 Å wide at the base and 12 Å deep, with a zinc(II) atom at the apex coordinated to three imidazole nitrogens from histidines and probably a water molecule to give a distorted tetrahedral coordination about the zinc. The zinc can be removed by chelating agents, and the resulting apoenzyme reacts with various metals such as cobalt(II), copper(II), cadmium(II), and nickel(II). The cobalt(II) derivative has about 50 percent of the catalytic activity of the zinc enzyme, and various spectroscopic measurements on the cobalt analogue have been helpful in elucidating the coordination chemistry.

The pH profile for the CO_2 hydration is consistent with the reactions in Scheme 8.3, with $pK_{a1} \approx 7$ and $pK_{a2} \approx 9$ (values vary somewhat with the isozyme and ionic medium). The assignment of these ionizations to particular subunits has been an area of controversy. Simple model systems would lead to the expectation that pK_{a1} is due to an imidazole nitrogen of histidine and that pK_{a2} is due to ionization of $(Im)_3Zn-OH_2^{2+}$. However, the opposite assignment is now generally favored. The problem with this is that many model zinc(II) complexes have values of $pK_a > 8$, but these models are five- or six- coordinate species[41,42] and it can be argued that four-coordinate zinc(II) in the enzyme will be more acidic than the models. Brown et al.[43] found that a trisimidazole chelate complex of cobalt(II) has a pK_a of 7.6, but the ionization is accompanied by a coordination change from six to either five or four based on changes in the electronic spectrum. The best model so far appears to be the trinitrogen chelate [12]aneN$_3$ of zinc(II), which Kimura et al.[44] have reported to have a pK_a of 7.3 and which mimics the aldehyde hydration and ester hydrolysis activity of carbonic anhydrase. Spectrophotometric titration[45] of the cobalt(II) derivative of human CA I gives two pK_a values of 6.9 and 8.7, and both ionizations have a similar effect on the electronic spectrum. The cadmium(II) derivatives have pK_{a1} in the range of 9.5 to 10.[46] Clearly, the metal ion is involved in the first ionization to give the active species, which is widely accepted to be $(Im)_3Zn-OH$. Therefore, the first step in the catalysis cycle can be viewed as formation of $(Im)_3Zn-OH$.

The next step is the hydration of the CO_2, which is taken to be attack of the zinc-bound OH$^-$ on CO_2, for which there are ample precedents[47] in the chemistry of carbonate complexes of inert metal ions. The only difference may be that the zinc can expand its coordination sphere from four to five by complexing with CO_2 in what is called the inner-sphere mechanism.[48] The outer-sphere alternative without any Zn–CO_2 interaction has been discussed by Lipscomb.[49] The theoretical work of Merz et al.[50] predicts the following transition state:

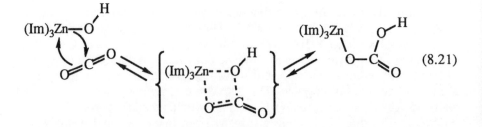

$$(8.21)$$

The third step is replacement of bicarbonate by water in the coordination sphere of the zinc. This should be a facile process because of the substitution lability of zinc(II). Finally, the $(Im)_3Zn$—OH_2 must be converted to $(Im)_3Zn$—OH by proton transfer to some base. This step is thought to be rate controlling under swamping steady-state conditions, because the rate shows a substantial deuterium isotope effect[51] and it is increased by more basic buffers.[52] In the enzyme, it is assumed that some base from the peptide acts as a proton acceptor and transfer agent. The base most often suggested is an imidazole from histidine(64), which is connected to the Zn—OH_2 by a water-bridged hydrogen bonding network.

Pocker and Janjíc[53] have suggested the mechanism in Figure 8.4, where B--- is the buffer species that may be complexed to the peptide. The need for a 180° flip of the imidazole could be questioned, since the acidity of the tautomeric forms of histamine are not greatly different.[54] Other proton relay networks have been suggested involving different peptide groups[55] or just water molecules in the active site cleft.[56] It has been noted that the replacement of histidine(64) by other amino acids in mutant enzymes reduces the activity only by 1.5 to 3 times. Pocker and Janjíc argued that these mutants may permit other sidechain amino acids to be active and may open the cavity to allow other basic buffer components to enter. These suspicions seem to be confirmed by the report of Tu et al.[57] that the alanine mutant has about 20 times lower activity in the absence of buffer, and the activity can be restored if imidazole or 1-methylimidazole buffers are used. Buffer effects on the rate of H_2O^{18}–CO_2 exchange by the less reactive human CA III also have been reported.[58]

Carbonic anhydrase is inhibited by a wide range of anions. Pocker and Diets[59] found that the inhibition is of the competitive type at pH 6.6 but is uncompetitive at pH 9.9. The uncompetitive inhibition implies that Zn—OH is not complexed by inhibitor, but zinc(II) in the enzyme substrate complex can expand its coordination number to five with the ligands being 3-imidazoles, substrate, and either water or inhibitor. Alkyl or aryl sulfonamides are strong inhibitors for carbonic anhydrase and bind as the anion, RSO_2NH^- or $ArSO_2NH^-$, respectively. Dugad et al.[60] found that p-fluorobenzenesulfonamide forms a bis complex, indicating that the zinc is quite capable of being five-coordinate. Liang and Lipscomb[61] have modeled the formation and bonding of acetamide and sulfonamide inhibitors.

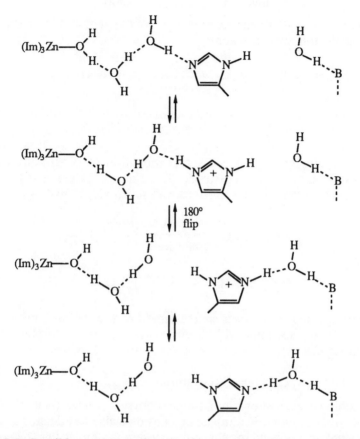

Figure 8.4. Possible proton transfer steps for carbonic anhydrase involving histidine(64) and buffer.

Bertini et al.[62] showed that nitrate binds to the EH_2 and EH forms of the cobalt(II)-substituted enzyme. They attribute this to the EH form having tautomers E(histidine-H)(Zn—OH) and E(histidine)(Zn—OH_2), with water replacement in the latter being the source of nitrate binding to EH. This maintains consistency with the standard interpretation that anions cannot compete with OH^- in the Zn(OH) species.

8.3 ENZYMIC REACTIONS OF DIOXYGEN

There are a number of metalloenzymes (EM) that bring about reactions of dioxygen and its derivatives, hydrogen peroxide and the superoxide ion. They may act as reversible oxygen carriers, such as hemoglobin and myoglobin in the following equation:

$$EM + O_2 \rightleftharpoons EM(O_2) \tag{8.22}$$

The enzyme may introduce either one or two oxygens into the substrate, as shown in the following reactions:

$$O_2 + SH + AH_2 \xrightarrow[\text{oxygenases}]{\text{Mono-}} SOH + H_2O + A \tag{8.23}$$

$$O_2 + SH \xrightarrow{\text{Dioxygenases}} SHO_2 \tag{8.24}$$

Other enzymes catalyze the disproportionation of superoxide ion and decomposition of hydrogen peroxide, as in Eqs. (8.25) and (8.26), in order to protect the system[63] from reactions with superoxide or hydroxy radicals.

$$2\,O_2^- + 2\,H^+ \xrightarrow[\text{dismutase}]{\text{Superoxide}} O_2 + H_2O_2 \tag{8.25}$$

$$H_2O_2 + AH_2 \xrightarrow{\text{Peroxidases}} H_2O + A \tag{8.26}$$

The oxidases and peroxidases are the subject of a recent, succinct review.[64] The following sections will discuss two specific examples of these metalloenzymes.

8.3.a   Oxygen Carriers: Myoglobin

Myoglobin consists of one iron–protoporphyrin complex (heme prosthetic group) with an imidazole nitrogen of a histidine also coordinated to the iron and a single peptide chain. Since this system is simpler than hemoglobin which has four heme units, much of the mechanistic work has been done on myoglobin. There are several sources of myoglobin which differ in the peptide composition but the work discussed in this section is on the most common source, sperm whale myoglobin, unless otherwise indicated. The thermodynamic and structural effects of dioxygen binding to myoglobin and hemoglobin have been reviewed recently.[65]

The structural features of the iron coordination in the iron(II) deoxy form and the dioxygen complex have been determined by Takano[66] and Phillips,[67] respectively, and are shown in Figure 8.5. The iron(II) in deoxymyoglobin is displaced by 0.42 Å out of the plane of the four porphyrin nitrogens toward the imidazole nitrogen. In the dioxygen complex, this displacement is reduced to 0.18 Å. The dioxygen is bonded in an end-on fashion with an Fe—O—O angle of 115°. Takano[68] reported the structure of the iron(III) complex (metmyoglobin) with axial water and imidazole ligands. These structural features also are found in the model system described by Collman and co-workers,[69] where the displacement of iron depends on the steric requirements of the imidazole ligand, being 0.086

Å with 2-methylimidazole and 0.03 Å with 1-methylimidazole.

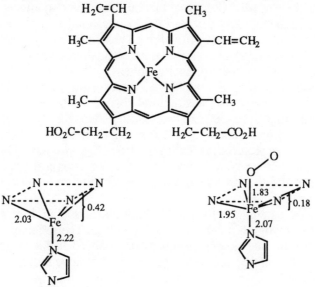

Figure 8.5. Iron protoporphyrin IX, the prosthetic group in myoglobin, and the coordination geometry of iron in myoglobin and oxymyoglobin.

In the development of model systems, a major problem has been that the simple iron(II) porphyrins form diiron μ-peroxo complexes, as shown in Eq. (8.27):

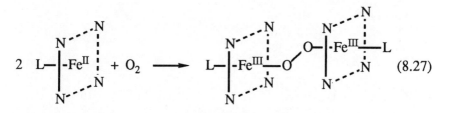

(8.27)

This can be overcome by introducing appropriate steric hindrance to dimer formation. This was achieved by Collman et al.[70] by using the steric bulk of *o*-pivalamidophenyl substituents on *meso*-tetraphenylporphyrin to prepare a model oxygen carrier. In the biological system, the peptide may serve a similar purpose.

In oxymyoglobin (MbO_2), the terminal oxygen is hydrogen bonded to an imidazole of a distal histidine(64). Mutants in which this histidine is replaced by other amino acids show a lower discrimination between O_2 and CO.[71] The dissociation rates of O_2 are increased by 50 to 1500 times, while the association rates for O_2 and CO increase 5 to 15 times. Resonance Raman studies[72] of MbCO indicate that this histidine also affects the Fe—C stretching frequency, and the pH effect indicates that protonation of the histidine occurs with a pK_a of ~4.5. These studies also revealed that removal of histidine(64) increases the autooxidation rate of MbO_2, so that the histidine is serving some protective role as well.

The complexation of myoglobin by dioxygen appears to be a simple process accompanied by some structural changes. However, the iron(II) is initially in the high-spin state and reacts with paramagnetic dioxygen to yield a diamagnetic product and a lot is happening electronically. The dioxygen complex can be pictured as $Fe^{II}(O_2)$, $Fe^{III}(O_2^-)$ or $Fe^{IV}(O_2^{2-})$. The $Fe^{III}(O_2^-)$ formalism is now favored as best describing the spectroscopic properties of the system. Therefore, the substitution is accompanied by electron transfer.

The kinetic parameters for the binding and dissociation of some small molecules to myoglobin are given in Table 8.1. The relatively small rate constants for the isocyanides have been rationalized as due to steric effects imposed by the peptide as these larger molecules move through the peptide

Table 8.1. Kinetic Parameters for Complexing of Myoglobin (Mb)

Reaction	k $(M^{-1} s^{-1})$	ΔH^* $(kJ\ mol^{-1})$	ΔS^* $(J\ mol^{-1}\ K^{-1})$	ΔV^* $(cm^3\ mol^{-1})$
Mb + O_2 [a]	2.5×10^7	26 [b]	−15	5.2
Mb + O_2 [c]	1.3×10^7	23.0	−30	7.8
Mb + O_2 [d]	2.4×10^7			4.6
Mb + CO [c]	3.8×10^5	17.1	−81.1	−8.9
Mb + CO [d]	6.7×10^5			−9.2
Mb + CO [e]	5.2×10^5			−10.0
Mb + MeNC	1.2×10^5 [f]			8.8 [e]
Mb + n-BuNC [f]	3.0×10^4			

[a] Projahn, H.-D.; Dreher, C.; van Eldik, R. *J. Am. Chem. Soc.* **1990**, *112*, 17; in 5 x 10^{-3} M Tris buffer, 0.1 M NaCl, pH 8.5, 25°C.

[b] Calculated from the dissociation rate and the equilibrium constant parameters.

[c] Hasinoff, B. B. *Biochemistry* **1974**, *13*, 3111; in 0.1 M phosphate buffer, pH 7.0, 25°C.

[d] Adachi, S.; Morishima, I. *J. Biol. Chem.* **1989**, *264*, 18896; in 0.1 M Tris buffer, pH 7.8, 20°C.

[e] Taube, D. J.; Projahn, H.-D.; van Eldik, R.; Magde, D.; Traylor, T. G. *J. Am. Chem. Soc.* **1990**, *112*, 6880; in 0.05 M bis-Tris buffer, 0.1 M NaCl, pH 7.0, 25°C.

[f] Rohlfs, R. J.; Olson, J. S.; Gibson. Q. H. *J. Biol. Chem.* **1988**, *263*, 1803; in 0.1 M phosphate buffer, pH 7, 20°C.

cleft toward the iron. However, the kinetic difference between O_2 and CO is more remarkable. The ΔV^* for CO binding is quite negative, whereas that for O_2 is positive, and the CO rate constant is smaller, largely because of the less favorable ΔS^*. The binding of CO gives a low-spin iron(II) complex with the iron 0.1 Å below the porphyrin plane,[73] compared to 0.18 Å in the O_2 complex. The O_2 is bent and hydrogen bonded to a distal histidine, whereas the CO is linear but still rather immobile according to ^{17}O nmr studies.[74] The binding of O_2 is independent of pH, and CO binding is faster at lower pH.[75]

It has been possible to obtain a detailed picture of the complexation process by using laser flash photolysis to dissociate the bound ligand and then watch its recombination or movement out of the active site pocket. Martin et al.[76] studied CO dissociation from several hemes using a 250-fs laser pulse at 307 nm and concluded that the high-spin iron(II) state is formed within 0.35 ps. Finsden at al.[77] followed the Raman spectrum of the Fe—N(imidazole) stretch, after dissociation of CO and O_2 from hemoglobin, and found that the deoxy form was produced within the 10-ns pulse length. This is rather surprising since low-spin to high-spin changes for octahedral iron(II) complexes[78,79] have rate constants of ~10^7 s^{-1}. Traylor and co-workers[80] observed the time evolution of difference spectra in the 400- to 500-nm range after flash photolysis and found one process on the nanosecond and another on the picosecond time scale. Olson et al.[81] have done similar experiments but using essentially nanosecond time scale detection. Both groups interpret the results in terms of a four-state model, described in Scheme 8.4, where B is called a geminate pair and the leaving group has moved further away and probably rotated somewhat in C.

Scheme 8.4

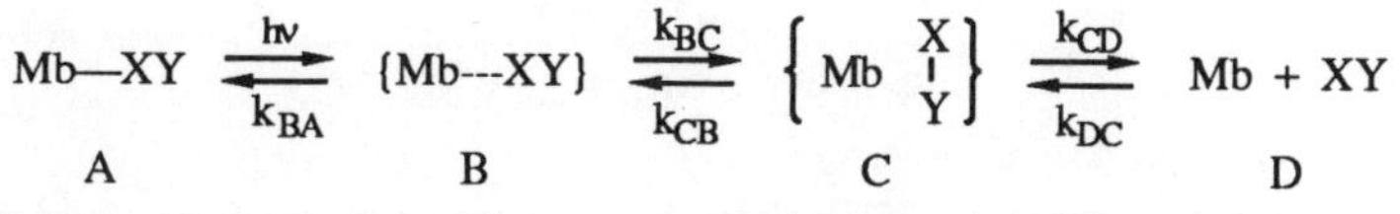

$$\underset{A}{Mb-XY} \underset{k_{BA}}{\overset{h\nu}{\rightleftharpoons}} \underset{B}{\{Mb\text{---}XY\}} \underset{k_{CB}}{\overset{k_{BC}}{\rightleftharpoons}} \underset{C}{\left\{ Mb \overset{X}{\underset{Y}{|}} \right\}} \underset{k_{DC}}{\overset{k_{CD}}{\rightleftharpoons}} \underset{D}{Mb + XY}$$

The kinetic observations combined with quantum yields have been used to determine rate constants for the various steps. Some data from both studies are given in Table 8.2. The results are at some variance, especially with regard to k_{BA} and k_{BC}. It should be noted that no transient time-dependent signals were observed for CO dissociation, although CO dissociation has a high quantum yield.[82] This could be explained by a larger k_{BC} for CO, but the value for O_2 is at the diffusion limit according to the data of Traylor et al, so that k_{BA} must be smaller to explain the observations. Since k_{BA} and k_{CB} do not show any correlation with the steric bulk of XY, Traylor and co-workers suggested that the slower overall complexation of isocyanides is due to changes in k_{DC} caused by steric effects as the ligand enters the protein pocket. This is consistent with a crystal structure of the ethyl isocyanide complex.[83]

Table 8.2. Rate Constants Following Flash Photolysis of Myoglobin–XY Complexes Analyzed According to Scheme 8.4

XY	k_{BA} (s^{-1})	k_{BC} (s^{-1})	k_{CB} (s^{-1})	k_{CD} (s^{-1})
O_2 [a]	1.4×10^{10}	1.5×10^{10}	4.6×10^6	8.8×10^6
O_2 [b]	4.9×10^8	1.2×10^8	8.5×10^6	1.4×10^7
CNMe [a]	1.3×10^{10}	4.0×10^{10}	7.8×10^7	5.9×10^6
CNMe [b]	2.4×10^7	4.9×10^6		
CN(n-Bu) [b]	1.6×10^7	1.4×10^7	8×10^5	1.2×10^6
CN(t-Bu) [a]	2.6×10^{10}	8.0×10^9	5.2×10^6	3.6×10^6

[a] Jongeward, K. A.; Magde, D.; Taube, D. J.; Marsters, J. C.; Traylor, T. G.; Sharma, V. S. *J. Am. Chem. Soc.* **1988**, *110*, 380; in 0.1 M Tris buffer, 0.1 M NaCl, pH 7, unspecified temperature.

[b] Rohlfs, R. J.; Olson, J. S.; Gibson. Q. H. *J. Biol. Chem.* **1988**, *263*, 1803; in 0.1 M phosphate buffer, pH 7, 20°C.

Recently, Magde and co-workers[84] have measured k_{CD} and its Arrhenius activation energy for several complexes of horse heart myoglobin and for the O_2 sperm whale myoglobin system. For the latter, they report $k_{CD} = 7.7 \times 10^6$ s^{-1} (25°C, 0.1 M Tris buffer, 0.1 M NaCl, pH 7.0) and $E_a = 7.4$ kcal mol^{-1}. The range of the E_a values for all the systems is 6 to 9 kcal mol^{-1}. They also suggest that a five-state model may be required, involving either two escape routes from the peptide pocket or a second bound site within the pocket.

The observed steady-state rate constant for Scheme 8.4 is given by

$$k_{obsd} = \frac{k_{BA}\,k_{CB}\,k_{DC}}{k_{BA}\,k_{CB} + k_{BA}\,k_{CD} + k_{BC}\,k_{CD}} \qquad (8.28)$$

If k_{BA} is unusually small for CO, then the first two terms in the denominator of Eq. (8.28) are small relative to the third term, and the equation simplifies to $k_{obsd} = k_{BA}(k_{CB}/k_{BC})\,(k_{DC}/k_{CD}) = k_{BA}K_{CB}K_{DC}$. This corresponds to step B to A being rate controlling and preceded by fast equilibria of D to C and C to B. For O_2, the first two terms in the denominator are somewhat larger than the third, so that k_{BA} approximately cancels in the numerator and denominator. Then, $k_{obsd} = k_{CB}(k_{DC}/k_{CD}) = k_{CB}K_{DC}$ because k_{CD} is larger than k_{CB}, and step C to B is rate-controlling. This would explain the kinetic differences between O_2 and CO binding.

Adachi and Morishima[85] have determined the pressure dependence for some of the steps in Scheme 8.4 for several myoglobins. The results, given in Table 8.3, show that changes in the peptide cause small changes in the rate constants but substantial differences in the activation volumes.

Table 8.3. Rate Constants (20°C) and Activation Volumes for the Reactions of O_2 and CO with Different Myoglobins[a]

	Sperm Whale	Horse	Dog
k_{CA} (s^{-1})	3.9×10^6	3.5×10^6	3.6×10^6
ΔV^*_{CA} (cm^3 mol^{-1})	3.5	-8.4	-17.8
k_{CD} (s^{-1})	5.6×10^6	6.2×10^6	7.4×10^6
ΔV^*_{CD} (cm^3 mol^{-1})	16.7	11.2	-2.1
k_{DC} (M^{-1} s^{-1})	5.7×10^7	7.9×10^7	9.1×10^7
ΔV^*_{DC} (cm^3 mol^{-1})	10.4	12.3	8.6
k_{obsd}[c] (M^{-1} s^{-1})	2.4×10^7	2.9×10^7	3.0×10^7
ΔV^*_{obsd}[c] (cm^3 mol^{-1})	4.6	3.8	0
k_{obsd} (CO)[c] (M^{-1} s^{-1})	6.7×10^6	6.8×10^6	8.5×10^6
ΔV^*_{obsd} (CO)[c] (cm^3 mol^{-1})	-9.2	-12.7	-18.8

[a] In 0.1 M Tris buffer, pH 7.8; values are for O_2 unless otherwise indicated.
[b] $k_{CA} = k_{BA}k_{CB}/(k_{BA} + k_{BC})$ from Scheme 8.4.
[c] Values from steady-state conditions.

The systems in Table 8.3 all have the histidine(64) residue that is involved in hydrogen bonding to the bound O_2, but they vary in the residues that may be affecting the size and structure of the pocket leading to the active site. Adachi and Morishima have given a more detailed rationalization of the effects of varying amino acid residues. Traylor[86] and co-workers subsequently reported activation volumes for k_{CD} with sperm whale myoglobin of 11.7, 12.6, and 9.1 cm^{-3} M^{-1} for CO, O_2, and MeNC, respectively. For the overall formation reaction, the more negative values of ΔV^*_{obsd}(CO) can be attributed to contractions occurring at the B to A stage, which follows the rate-limiting step for O_2 binding and therefore does not influence ΔV^*_{obsd}(O_2).

8.3.b Cytochrome P-450

There is a large and wide-ranging family of monooxygenase enzymes, called cytochrome P-450s, whose chemistry and biochemistry are the subjects of a recent book.[87] They take their name from their characteristic absorbance at 450 nm. They all contain a heme prosthetic group and catalyze the reactions of O_2 and a reducing agent to bring about transformations, such as those in Eqs. (8.29) to (8.32). The oxygen in the substrate is derived from the O_2. These enzymes are rather unspecific in their reactivity, and these are only a few examples of the many reactions catalyzed by P-450 enzymes.

$$R\text{-}CH_3 + O_2 + 2\,H^+ + 2\,e^- \xrightarrow{\text{P-450}} R\text{-}CH_2OH + H_2O \qquad (8.29)$$

$$R-CH=CH_2 + O_2 + 2\,H^+ + 2\,e^- \xrightarrow{\text{P-450}} \underset{RHC\!\!-\!\!-\!\!CH_2}{\overset{O}{\triangle}} + H_2O \qquad (8.30)$$

$$\text{(benzene)} + O_2 + 2\,H^+ + 2\,e^- \xrightarrow{\text{P-450}} \text{(phenol)}-OH + H_2O \qquad (8.31)$$

(8.32)

Reaction (8.32) represents the transformation of camphor to the exo-alcohol by P-450-CAM, one of the best characterized of these enzymes. The crystal structure of P-450-CAM and its camphor adduct have been determined by Poulos and co-workers.[88,89] The solution state coordination of iron has been characterized by Dawson and co-workers[90,91] using extended x-ray absorption fine structure (EXAFS). The resting state prosthetic group has a six-coordinate low-spin iron(III) heme with axial S and O ligands. The S is derived from cysteine, and Poulos has argued from bond lengths (Fe—S = 2.2 Å) that it is a thiolate cys-S⁻—Fe linkage. The O is either from an OH⁻ or an OH$_2$. The iron(III) is displaced 0.29 Å from the plane of the porphyrin nitrogens toward the S⁻. The pocket leading to the active site contains five hydrogen-bonded water molecules and is lined with hydrophobic amino acid groups with no acid or base residues. If camphor is added to P-450-CAM under anaerobic conditions, then camphor displaces water and moves into the active site pocket. This complex contains five-coordinate high-spin iron(III), with the iron 0.43 Å out of the porphyrin plane and an essentially unchanged Fe—S bond length. The camphor is poised over the heme, with the C to be hydroxylated closest to the iron and the C=O of camphor hydrogen bonded to an -OH of a tyrosine. An EXAFS study by Dawson and co-workers of the low-spin iron(II)–camphor–O$_2$ adduct, prepared at –40°C in water–ethylene glycol, indicates bond lengths very similar to those of a model thiolate–Fe^{II}(porphyrin)–O$_2$ complex.[92]

Since the operation of P-450 requires the enzyme plus three reagents (substrate, O$_2$, and reducing agent), it is possible to control the reaction conditions in order to identify intermediates and to establish the reaction pathway.[93] Current evidence is consistent with the sequence in Figure 8.6. A high-spin iron(III) complex forms when substrate (R–C–H) enters the active site pocket, due to small changes in the peptide conformation or loss of the water ligand. This form is more easily reducible by 0.12 V[94] than the resting enzyme and favors the reduction to high spin-iron(II), which then forms the O$_2$ complex similar to myoglobin.

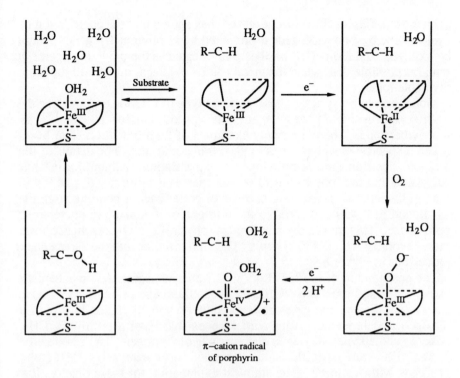

Figure 8.6. A catalytic cycle for alkane hydroxylation by cytochrome P-450 in which the first four intermediates have been identified and the π-cation radical is speculative.

Precedent for the π-cation radical Fe=O species comes from the more stable and well characterized intermediate HRP I[95,96] of horseradish peroxidase. Furthermore, it has been shown that the iron(III) resting state of P-450 reacts with iodosylbenzene to produce a species that hydroxylates alkanes. This can be understood if oxygen transfer from iodosylbenzene proceeds, as shown in Eq. (8.33), to give the π-cation radical Fe=O species:

$$\text{PhIO} + (\text{porphyrin})\text{Fe}^{III} \longrightarrow \text{PhI} + (\text{porphyrin}\cdot^{+})\text{Fe}^{IV}=\text{O} \qquad (8.33)$$

This synthetic version causes the same hydroxylations as the natural system. However, ^{18}O exchange with $^{18}\text{OH}_2$ is observed for the iodosylbenzene product while the natural system shows no exchange.[97] Groves and et al.[98] found that cytochrome P-450$_{LM2}$ with its reductase gives exchange of the trans 1-H of propylene with solvent D_2O, but the enzyme with iodosylbenzene gave no exchange. The iodosylbenzene reactions may be different because the substrate enters the active site only after the iodosylbenzene enters and phenyl iodide leaves. These reactions may be even more different since recent work[99] has shown that the epoxidation of cyclohexene by iodosylbenzene is promoted by aluminum(III), which cannot be oxidized, and iodine(III) has been implicated as the oxidizing agent. It remains possible that, when the metal can be oxidized, this is still the more favorable route.

Model porphyrin systems show the same type of reaction and catalyze the hydroxylation of alkanes[100] and the epoxidation of alkenes[101] by iodosylbenzene. There are many examples of high oxidation state M=O (oxene) species[102] which act as oxygen transfer agents. The details of the oxygen insertion step remain open to speculation. Champion[103] has suggested that the (porphyrin•+)Fe^{IV}=O species is more reactive in P-450 than in horseradish peroxidase because of better back π bonding from the -S$^-$ ligand in P-450 compared to the nitrogen of imidazole in horseradish peroxidase. This makes the reactive (porphyrin)Fe^{IV}(O•) resonance form more favorable with P-450 systems. Then the reaction with the alkane could proceed as in Figure 8.7.

The epoxidation of olefins [Eq. (8.30)] presents some new mechanistic possibilities and problems. Ortiz de Montellano and co-workers[104] have shown that the enzyme gives both epoxidation and alkylation of a pyrrole nitrogen, the latter occurring with alkenes that have a terminal =CH$_2$. Alkenes and alkynes alkylate a different pyrrole nitrogen. The epoxidation seems to be stereospecific, since the reaction of *trans*-1-[1-^{2}H]octene proceeds with retention. The simplest explanation for these observations would seem to be that the peptide controls the orientation of the substrate over the iron and that the reactions proceed as in Figure 8.8.

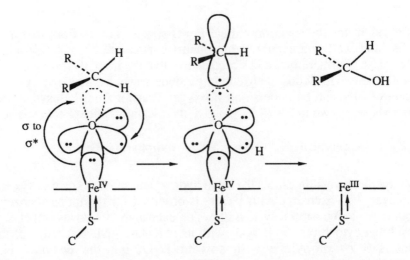

Figure 8.7. Possible product formation steps for P-450 hydroxylation starting from the (porphyrin)Fe^{IV}(O•) resonance form favored by back π bonding from S$^-$.

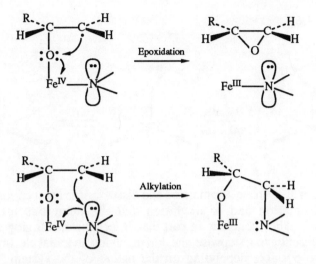

Figure 8.8. Possible routes to epoxide and N-alkylation products in the reaction of alkenes with P-450 systems.

A concerted mechanism such as that in Figure 8.8 would not predict the deuterium exchange results of Groves and co-workers, and models involving organo-iron species have been invoked[105,106] based on known model chemistry. The deuterium exchange was originally explained as proceeding through the metalocycle in Scheme 8.5.

Scheme 8.5

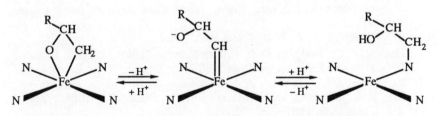

Dolphin et al.[107] have suggested, based on model studies with tetra-(2,6-dichlorophenyl)porphyrin, that a ß-hydroxyalkyl species can be formed from the N-alkyl derivative under reducing conditions, as shown in Scheme 8.6. This provides an alkyl derivative that gives proton exchange through deprotonation to the carbene. Mansuy and co-workers[108] have reported efficient preparations of iron *meso*-(tetraphenyl)porphyrin carbenes such as $(N)_4Fe^{IV}(=C(CH_3)COPh)$. Sterically crowded alkenes are reactive with both P-450 and model systems, yet they do not show N-alkylation and are unlikely to form carbenes. Therefore, the latter processes appear to be side reactions of the main epoxidation process.

Scheme 8.6

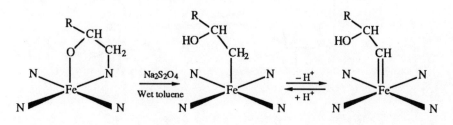

Collman et al.[109] have summarized and discussed the competition between epoxide formation and N-alkylation and the numerous mechanistic possibilities. They conclude in part that "Mechanisms ranging from fully concerted reaction to a stepwise reaction involving an initial electron transfer may all be possible depending on the nature of the system." Such a conclusion is encouraging for those seeking challenging mechanistic problems, but discouraging to those hoping for a unified mechanism to explain all observations.

References

1. Williams, R. J. P. *Coord. Chem. Rev.* **1987**, *79*, 175.
2. *Metal Ions in Biological Systems*; Sigel, H., Ed.; Marcel Dekker: New York; *Advances in Inorganic Biochemistry*; Eichorn, G. L.; Marzilli, L. G., Eds.; Elsevier: New York; *Advances in Inorganic and Bioinorganic Mechanisms*; Sykes, A. G., Ed.; Academic Press: New York; *Metal Ions in Biology*; Spiro, T. G., Ed.; Wiley-Interscience: New York.
3. *J. Inorg. Biochem.*; *Bioinorg. Chem.*; *Inorg. Chim. Acta*; *Bioinorg. Chem. Art. Lett.*
4. *B12*; Dolphin, D., Ed.; Wiley: New York, 1982.
5. Lenhert, P. G.; Hodgkin, D. C. *Nature (London)* **1961**, *192*, 937; Lenhert, P. G. *Proc. R. Soc. London* **1968**, *A303*, 45; Savage, H. F.; Lindley, P. F.; Finney, J. L.; Timmins, P. A. *Acta Crystallogr.* **1987**, *B43*, 296.
6. Pett, V. B.; Liebman, M. N.; Murray-Rust, P.; Prasad, K.; Glusker, J. P. *J. Am. Chem. Soc.* **1987**, *109*, 3207.
7. Rossi, M.; Glusker, J. P.; Randaccio, L.; Summers, M. F.; Toscano, P. J.; Marzilli, L. G. *J. Am. Chem. Soc.* **1985**, *107*, 1729.
8. Kräutler, B.; Keller, W.; Kratky, C. *J. Am. Chem. Soc.* **1989**, *111*, 8936.
9. Sagi, I.; Wirt, M. D.; Chen, E.; Frisbie, S.; Chance, M. R. *J. Am. Chem. Soc.* **1990**, *112*, 8639.
10. Shrauzer, G. N. *Acc. Chem. Res.* **1968**, *1*, 97.
11. Finke, R. G.; Smith, B. L.; McKenna, W.; Christian, P. A. *Inorg. Chem.* **1981**, *20*, 687.
12. Costa, G. *Coord. Chem. Rev.* **1972**, *8*, 63.
13. Rétey, J.; Umani-Ronchi, A.; Seibl, J.; Arigoni, D. *Experientia* **1966**, *22*, 502.

14. Babior, B. M.; Moss, T. A.; Orme-Johnson, W. H.; Beinert, H. *J. Biol. Chem.* **1974**, *249*, 4537.
15. Valinsky, J. E.; Abeles, J. E.; Fee, J. A. *J. Am. Chem. Soc.* **1974**, *96*, 4709.
16. Babior, B. M. *Acc. Chem. Res.* **1975**, *8*, 376, and references therein.
17. Pratt, J. M. In B_{12}; Dolphin, D. Ed.; Wiley: New York, 1982; Vol. 1.
18. Halpern, J. *Science* **1985**, *227*, 869.
19. Finke, R. G.; Schiraldi, D. A.; Mayer, B. J. *Coord. Chem. Rev.* **1984**, *54*, 1.
20. Halpern, J.; Ng, F. T. T.; Rempel, G. L. *J. Am. Chem. Soc.* **1979**, *101*, 7124.
21. Endicott, J. F.; Ferraudi, G. J. *J. Am. Chem. Soc.* **1977**, *99*, 243.
22. Nie, S.; Marzilli, P. A.; Marzilli, L. G.; Yu, N.-T. *J. Am. Chem. Soc.* **1990**, *112*, 6084.
23. Halpern, J. *Polyhedron* **1988**, *7*, 1483.
24. Koenig, T. W.; Hay, B. P.; Finke, R. G. *Polyhedron* **1988**, *7*, 1499.
25. Daikh, B. E.; Huthchison, J. E.; Gray, N. E.; Smith, B. L.; Weakley, J. R.; Finke, R. G. *J. Am. Chem. Soc.* **1990**, *112*, 7830.
26. Finke, R. G.; Hay, B. P. *Inorg. Chem.* **1984**, *23*, 3043; Ibid. **1985**, *24*, 1278.
27. Halpern, J.; Kim, S. H.; Leung, T. W. *J. Am. Chem. Soc.* **1984**, *106*, 8317; Ibid. **1985**, *107*, 2199.
28. Hay, B. P.; Finke, R. G. *J. Am. Chem. Soc.* **1986**, *108*, 4820.
29. Martin, B. D.; Finke, R. G. *J. Am. Chem. Soc.* **1990**, *112*, 2419.
30. Wollowitz, S.; Halpern, J. *J. Am. Chem. Soc.* **1988**, *110*, 3112.
31. Choi, S.-C.; Dowd, P. *J. Am. Chem. Soc.* **1989**, *111*, 2313, and references therein.
32. Murakami, Y.; Hisaeda, Y.; Ozaki, T.; Ohno, T.; Fan, S.-D.; Matsuda, Y. *Chem. Lett.* **1988**, 839.
33. Dowd, P.; Choi, S.-C.; Duak, F.; Kaufman, C. *Tetrahedron* **1988**, *44*, 2137.
34. Murakami, Y.; Hisaeda, Y.; Kikuchi, J.; Ohno, T.; Suzuki, M.; Matsuda, Y.; Matsura, T. *J. Chem. Soc., Perkin Trans. 2* **1988**, 1237.
35. Finke, R. G.; McKenna, W. P.; Schiraldi, D. A.; Smith, B. L.; Pierpont, C. *J. Am. Chem. Soc.* **1983**, *105*, 7592; Finke, R. G.; Schiraldi, D. A. *J. Am. Chem. Soc.* **1983**, *105*, 7605.
36. Silverman, R. B.; Dolphin, D. *J. Am. Chem. Soc.* **1976**, *98*, 4633.
37. Arigoni, D. In *Vitamin B_{12}, Proceedings of the Third European Symposium on Vitamin B_{12} and Intrinsic Cofactor*; Zagalak, B.; Friedrich, W., Eds.; Walter de Gruyter: Berlin, 1979; p. 389.
38. Chen, E.; Chance, M. R. *J. Biol. Chem.* **1990**, *265*, 12987.
39. *Zinc Enzymes, Prog. Inorg. Biochem. Biophys.*; Bertini, I.; Luchinat, C.; Maret, W.; Zeppezauer, M., Eds.; Birkhauser: Boston, 1986, Vol. 1.
40. Silverman, D. A.; Lindskog, S. *Acc. Chem. Res.* **1988**, *21*, 30.
41. Woolley, P. *Nature* (London) **1975**, *258*, 677.
42. Chaberek, S.; Courtney, R. C.; Martell, A. E. *J. Am. Chem. Soc.* **1952**, *74*, 5057.
43. Brown, R. S.; Salmon, D.; Curtis, N. J.; Kusuma, S. *J. Am. Chem. Soc.* **1982**, *104*, 3188.

44. Kimura, E.; Shiota, T.; Koike, T.; Shiro, M.; Kodama, M. *J. Am. Chem. Soc.* **1990**, *112*, 5805.
45. Bertini, I.; Dei, A.; Luchinat, C.; Monnanni, R. *Inorg. Chem.* **1985**, *24*, 301.
46. Bauer, R.; Limkilde, P.; Johansen, J. T. *Biochemistry* **1976**, *15*, 334; Tibell, L.; Lindskog, S. *Biochim. Biophys. Acta* **1984**, *778*, 110.
47. Palmer, D. A.; van Eldik, R. *Chem. Rev.* **1983**, *83*, 651.
48. Pullman, A. *Ann. N. Y. Acad. Sci.* **1981**, *367*, 340, and references therein.
49. Lipscomb, W. N. *Ann. Rev. Biochem.* **1983**, *52*, 17.
50. Merz, K. M., Jr.; Hoffmann, R.; Dewar, M. J. S. *J. Am. Chem. Soc.* **1989**, *111*, 5636.
51. Pocker, Y.; Bjorkquist, D. W. *Biochemistry* **1977**, *16*, 5698.
52. Pocker, Y.; Janji´c, N.; Miao, C. H. *Prog. Inorg. Biochem. Biophys.* **1986**, *1*, 341.
53. Pocker, Y.; Janji´c, N. *J. Am. Chem. Soc.* **1989**, *111*, 731.
54. Roberts, J. D.; Yu, C.; Flanagan, C.; Birdseye, T. *J. Am. Chem. Soc.* **1982**, *104*, 3945.
55. Kannan, K. K.; Ramanadham, M.; Jones, T. A. *Ann. N. Y. Acad. Sci.* **1984**, *429*, 49.
56. Vedani, A.; Huhta, D. W.; Jacober, S. P. *J. Am. Chem. Soc.* **1989**, *111*, 4075.
57. Tu, C.; Silverman, D. N.; Forsman, C.; Jonsson, B.-H.; Lindskog, S. *Biochemistry* **1989**, *28*, 7913.
58. Tu, C.; Paranawithana, S. R.; Jewell, D. A.; Tanhauser, S. M.; LoGrasso, P. V.; Wynns, G. C.; Laipis, P. J.; Silverman, D. A. *Biochemistry* **1990**, *29*, 6400.
59. Pocker, Y.; Diets, T. L. *J. Am. Chem. Soc.* **1982**, *104*, 2424.
60. Dugad, l. B.; Cooley, C. R.; Gerig, J. J. *Biochemistry* **1989**, *28*, 3955.
61. Liang, J.-Y.; Lipscomb, W. N. *Biochemistry* **1989**, *28*, 9724.
62. Bertini, I.; Dei, A.; Luchinat, C.; Monnanni, R. *Prog. Inorg. Biochem. Biophys.* **1986**, *1*, 371.
63. Imlay, J. A.; Linn, S. *Science* **1988**, *240*, 1302.
64. Dawson, J. H. *Science*, **1988**, *240*, 433.
65. Perutz, M. F.; Fermi, G.; Luisi, B.; Shaanan, B.; Liddington, R. C. *Acc. Chem. Res.* **1987**, *20*, 309.
66. Takano, T. *J. Mol. Biol.* **1977**, *110*, 569.
67. Phillips, S. E. V. *J. Mol. Biol.* **1980**, *142*, 531.
68. Takano, T. *J. Mol. Biol.* **1977**, *110*, 537.
69. Jameson, G. B.; Molinaro, F. S.; Ibers, J. A.; Collman, J. P.; Brauman, J. I.; Rose, E.; Suslick, K. S. *J. Am. Chem. Soc.* **1980**, *102*, 3224.
70. Collman, J. P.; Gagne, R. R.; Halbert, T. R.; Marchon, J.-C.; Reed, C. A. *J. Am. Chem. Soc.* **1973**, *95*, 7868.
71. Springer, B. A.; Egeberg, K. D.; Sligar, S. G.; Rohlfs, R. J.; Mathews, A. J.; Olson, J. S. *J. Biol. Chem.* **1989**, *264*, 3057.
72. Morikis, D.; Champion, P. M.; Springer, B. A.; Sligar, S. G. *Biochemistry* **1989**, *28*, 4791; Ramsden, J.; Spiro, T. G. *Biochemistry* **1989**, *28*, 3125.
73. Norvell, J. C.; Nunes, A. C.; Schoenborn, B. P. *Science* **1975**, *190*, 569.
74. Lee, C. L.; Oldfield, E. *J. Am. Chem. Soc.* **1989**, *111*, 1584.
75. Coletta, M.; Ascenzi, P.; Traylor, T. G.; Brunori, M. *J. Biol. Chem.* **1985**, *260*, 4151.

76. Martin, J. L.; Migus, A.; Poyart, C.; Lecarpentier, Y.; Astier, R.; Antonetti, A. *Proc. Natl. Acad. Sci. U.S.A.* **1983**, *80*, 173.
77. Finsden, E. W.; Friedman, J. M.; Ondrias, M. R.; Simon, S. R. *Science*, **1985**, *229*, 661.
78. Beatie, J. K.; Binstead, R. A.; West, R. J. *J. Am. Chem. Soc.* **1978**, *100*, 3046; McGarvey, J. J.; Lawthers, I.; Heremans, K.; Toftlund, H. *Inorg. Chem.* **1990**, *29*, 252.
79. Beattie, J. K. *Adv. Inorg. Chem.* **1988**, *32*, 2.
80. Jongeward, K. A.; Magde, D.; Taube, D. J.; Marsters, J. C.; Traylor, T. G.; Sharma, V. S. *J. Am. Chem. Soc.* **1988**, *110*, 380.
81. Olson, J. S.; Rohlfs, R. J.; Gibson. Q. H. *J. Biol. Chem.* **1987**, *262*, 12930.
82. Gibson, Q. H.; Olson, J. S.; McKinnie, R. E.; Rohlfs, R. J. *J. Biol. Chem.* **1986**, *261*, 10228.
83. Johnson, K. A.; Olson, J. S.; Phillips, G. N., Jr.; *J. Mol. Biol.* **1989**, *207*, 459.
84. Chatfield, M. D.; Walda, K. N.; Magde, D. *J. Am. Chem. Soc.* **1990**, *112*, 4680.
85. Adachi, S.; Morishima, I. *J. Biol. Chem.* **1989**, *264*, 18896.
86. Taube, D. J.; Projahn, H.-D.; van Eldik, R.; Magde, D.; Traylor, T. G. *J. Am. Chem. Soc.* **1990**, *112*, 6880
87. *Cytochrome P-450: Structure, Mechanism and Biochemistry*; Ortiz de Montellano, P. R., Ed.; Plenum Press: New York, 1986.
88. Poulos, T. L.; Howard, A. J. *Biochemistry* **1987**, *26*, 8165; Poulos, T. L.; Finzel, A. J.; Howard, J. *J. Mol. Biol.* **1987**, *195*, 687.
89. Poulos, T. L. *Adv. Inorg. Biochem.* **1987**, *7*, 1.
90. Dawson, J. H.; Kau, L.-S.; Penner-Hahn, J. E.; Sono, M.; Eble, K. S.; Bruce, G. S.; Hager, L. P.; Hodgson, K. O. *J. Am. Chem. Soc.* **1986**, *108*, 8114.
91. Dawson, J. H.; Sono, M. *Chem. Rev.* **1987**, *87*, 1255.
92. Ricard, L.; Schappacher, M.; Weiss, R.; Montiel-Montoya, R.; Bill, E.; Gonser, U.; Trautwein, A. *Nouv. J. Chim.* **1983**, *7*, 405.
93. Gunsalus, I. C.; Meeks, J. R.; Lipscomb, J. D.; Debnner, P.; Munck, E. In *Molecular Mechanisms of Oxygen Activation*; Hayaishi, O., Ed.; Academic Press: New York, 1974, p. 559.
94. Sligar, S. G.; Gunsalus, I. C. *Proc. Nat. Acad. Sci. U. S. A.* **1976**, *73*, 1078.
95. Dunford, H. B. *Adv. Inorg. Biochem.* **1982**, *4*, 41.
96. Penner-Hahn, J. E.; Eble, K. S.; McMurry, T. J.; Renner, M.; Balch, A. L.; Groves, J. T.; Dawson, J. H.; Hodgson, K. O. *J. Am. Chem. Soc.* **1986**, *108*, 7819.
97. McDonald, T. L.; Burka, L. T.; Wright, S. T.; Guengerich, F. P. *Biochem. Biophys. Res. Commun.* **1982**, *104*, 620.
98. Groves, J. T.; Avaria-Neisser, G. E.; Fish, K. M.; Imachi, M.; Kuczkowsli, R. L. *J. Am. Chem. Soc.* **1986**, *108*, 3837.
99. Yang, Y.; Diedrich, F.; Valentine, J. S. *J. Am. Chem. Soc.* **1990**, *112*, 7826.
100. Groves, J. T.; Nemo, T. E.; Myers, R. S. *J. Am. Chem. Soc.* **1979**, *101*, 1032; Chang, C. K.; Kuo, M.-S. *J. Am. Chem. Soc.* **1979**, *101*, 3413.

101. Ostovi`c, D.; Bruice, T. C. *J. Am. Chem. Soc.* **1989**, *111*, 6511.
102. Holm, R. H. *Chem. Rev.* **1987**, *87*, 1401.
103. Champion, P. M. *J. Am. Chem. Soc.* **1989**, *111*, 3433.
104. Ortiz de Montellano, P. R.; Mangold, B. L. K.; Wheeler, C.; Kunze, K. L.; Reich, N. O. *J. Biol. Chem.* **1983**, *258*, 4208; Kunze, K. L.; Mangold, B. L. K.; Wheeler, C.; Beilan, H. S.; Ortiz de Montellano, P. R. *J. Biol. Chem.* **1983**, *258*, 4202.
105. Mansuy, D. *Pure Appl. Chem.* **1987**, *59*, 759.
106. Brothers, P. J.; Collman, J. P. *Acc. Chem. Res.* **1986**, *19*, 209.
107. Dolphin, D.; Matsumoto, A.; Shortman, C. *J. Am. Chem. Soc.* **1989**, *111*, 411.
108. Artaud, I.; Gregoire, N.; Battioni, J.-P.; Dupre, D.; Mansuy, D. *J. Am. Chem. Soc.* **1988**, *110*, 8714.
109. Collman, J. P.; Hampton, P. D.; Brauman, J. I. *J. Am. Chem. Soc.* **1990**, *112*, 2986.

Problems

CHAPTER 1

1. The following reaction is studied in *t*-butanol at 35°C by measuring the integrated peak intensity (I) of the ^{1}H nmr of the product.

$$cis\text{-}(H_3N)_2M(Cl)_2 + Br^- \longrightarrow cis\text{-}(H_3N)_2M(Cl)(Br) + Cl^-$$

The following table gives the experimental conditions for several kinetic runs and the values of I at various times.

(a) Determine the experimental rate constant for each of the runs.

(b) Plot the experimental rate constants versus the bromide ion concentration, and calculate the specific rate constant for the reaction, if possible.

| $(H_3N)_2M(Cl)_2$, (M) | 3.0×10^{-3} | | 5.0×10^{-3} | | 8.0×10^{-3} | |
Br^-, (M)	0.060		0.085		0.110	
	Time (s)	I	Time (s)	I	Time (s)	I
	50	0.30	50	0.68	50	1.4
	100	0.58	100	1.30	100	2.56
	150	0.80	150	1.80	150	3.50
	200	1.02	200	2.24	200	4.30
	300	1.40	300	2.94	250	4.94
	400	1.70	400	3.48	300	5.48
	600	2.16	500	3.88	350	5.92
	800	2.44	600	4.16	400	6.28
	1000	2.64	800	4.54	450	6.60
	1300	2.80	1000	4.74	500	6.82
	1600	2.90	1300	4.90	600	7.21
	2000	2.96	1600	4.96	800	7.64
	3500	3.00	2500	5.00	2000	8.00

2. (a) Develop an equation that can be used to determine the experimental rate constant from measurements of absorbance versus time for the following system.

$$A \; \underset{k_r}{\overset{k_f}{\rightleftharpoons}} \; B$$

Assume that both A and B are absorbing at the observation wavelength and that both species obey Beer's law and have molar absorbancies ε_A and ε_B, respectively.

(b) Show how the final absorbance can be used to determine the equilibrium constant for the reaction if ε_A and ε_B are known.

3. Develop an expression that could be used to determine the second-order rate constant for the following system. (See section 1.2.d.)

$$A + B \; \underset{k_{-2}}{\overset{k_2}{\rightleftharpoons}} \; 2\,C$$

4. (a) There have been two studies of the following isomerization reaction.

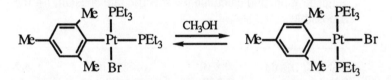

The results of the studies of the variation of the rate constant with temperature are given in the following table. Use each set of data to calculate the activation enthalpy and entropy for the reaction.

t (°C)	$10^4 \times k_{exp}$ (s^{-1})[a]	t (°C)	$10^4 \times k_{exp}$ (s^{-1})[b]
25	1.0	17.0	0.45
30	1.7	23.5	0.95
35	2.7	30.0	1.9
40	4.0	36.2	3.9
45	6.0	44.6	9.4

[a] Romeo, R.; Minniti, D.; Trozzi, M. *Inorg. Chem.* **1976**, *15*, 1134, using 5×10^{-5} M Pt(II) in 0.01 M LiClO$_4$.
[b] van Eldik, R.; Palmer, D. A.; Kelm, H. *Inorg. Chem.* **1979**, *18*, 572, using 5×10^{-4} M Pt(II) without added inert salt.

(b) Romeo et al. found that the rate of the reaction in part (a) is affected by the bromide ion concentration. The results at 30°C are given in the following table. Devise an empirical expression that will describe the dependence of k_{exp} on $[Br^-]$.

$10^4 \times [Br^-]$ (M)	0.0	2.0	4.0	6.0	8.0
$10^4 \times k_{exp}$ (s^{-1})	2.13	1.05	0.69	0.53	0.43

(c) van Eldik et al. also studied the pressure dependence of the following substitution reaction.

$$\underset{\substack{| \\ Br}}{\overset{\substack{PEt_3 \\ |}}{R-Pt-PEt_3}} + S=C(NH_2)_2 \underset{CH_3OH}{\rightleftharpoons} \left(\underset{\substack{| \\ S=C(NH_2)_2}}{\overset{\substack{PEt_3 \\ |}}{R-Pt-PEt_3}} \right)^+ + Br^-$$

$R \equiv 2,4,6$ trimethylbenzene

Under pseudo-first-order conditions (1.0×10^{-4} M Pt(II), 0.01 to 0.1 M S=C(NH$_2$)$_2$) they found the following two-term expression for the rate constant.

$$k_{exp} = k_1 + k_2 [S=C(NH_2)_2]$$

Use the pressure dependencies of k_1 and k_2 in the following table to calculate the volume of activation for each rate constant.

Pressure (bar)	$10^4 \times k_1$ (s^{-1})	$10^3 \times k_2$ (s^{-1})
1	2.25	3.25
250	2.53	4.27
500	3.08	4.82
750	3.72	5.54
1000	4.36	5.56

CHAPTER 2

1. Develop the expression for the pseudo-first-order rate constant for the following mechanism. Assume that K_i is a rapidly maintained equilibrium and that $[SO_4^{2-}] \gg [Co(III)]_{total}$.

$$(NH_3)_5CoOH_2^{3+} + SO_4^{2-} \underset{}{\overset{K_i}{\rightleftharpoons}} [(NH_3)_5CoOH_2 \cdot SO_4]^+ \quad \begin{array}{c} \text{Ion} \\ \text{pair} \end{array}$$

$$\downarrow k_1$$

$$(NH_3)_5CoOSO_3^+ + H_2O$$

2. Develop the expression for the pseudo-first-order rate constant for the following mechanism. Assume that [Y] and [X] >> [M]$_{total}$, the first step is a rapidly maintained equilibrium, and a steady state for the intermediate {(M)•Y}.

$$M{-}X + Y \overset{K}{\rightleftharpoons} [(M{-}X)\cdot Y] \underset{k_2}{\overset{k_1}{\rightleftharpoons}} \{(M)\cdot Y\} + X$$

$$\downarrow k_3$$

$$M{-}Y$$

3. The mechanism shown below might be suggested as an explanation for the effect of bromide ion on the isomerization reaction in problem 4(b) of Chapter 1.

$$\begin{array}{ccc} L & L & L \\ | & | & | \\ L{-}Pt{-}Br & \overset{+CH_3OH}{\rightleftharpoons} \quad L{-}Pt{-}OHCH_3 & \overset{+Br^-}{\rightleftharpoons} \quad R{-}Pt{-}Br \\ | & | & | \\ R & R & L \end{array}$$

$$+ Br^- \qquad\qquad + CH_3OH$$

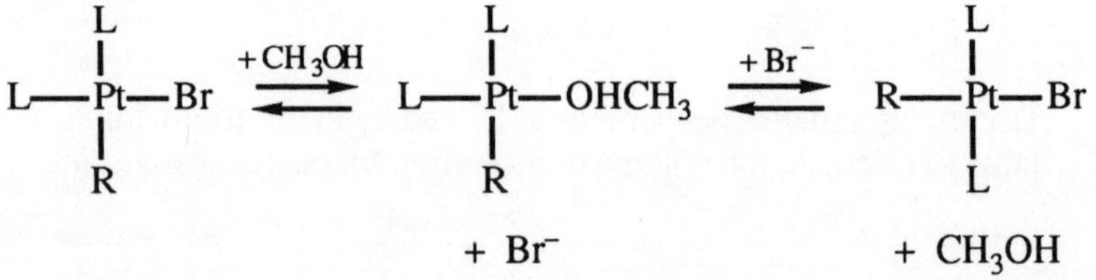

(a) Develop the rate law for this mechanism by assuming a steady state for the intermediate methanol complex.

(b) Does the predicted pseudo-first-order rate constant have a dependence on [Br$^-$] consistent with the data in problem 4(b) of Chapter 1?

(c) Is the mechanism as shown consistent with microscopic reversibility?

CHAPTER 3

1. The substitution reactions of (tetraphenylporphinato)chromium(III) chloride have been studied with several different leaving and entering groups. The kinetic results have been interpreted in terms of a limiting **D** mechanism as described by the following sequence.

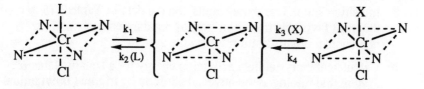

Some of the kinetic results (25°C in toluene) are summarized in the following table.

L	X	k_1 (s^{-1})[a]	k_3/k_2 [a]	log K[a]
PPh$_3$	MeIm[b]	4.6	1.03 x 10^3	4.8
P(OPr)$_3$	MeIm[b]	95	24	4.1
P(C$_2$H$_4$CN)$_3$	MeIm[b]	80	~2 x 10^2	(5.3)[c]
py	MeIm[b]	5.0	1.7	2.5
PPh$_3$	py	3.6		2.6

[a] O'Brien, P.; Sweigart, D. A. *Inorg. Chem.* **1982**, *21*, 2094.
[b] MeIm is N-methylimidazole.
[c] Calculated from data in the table.

(a) Use other data in the table to predict k_3/k_2 for the last system listed.

(b) Determine the order of nucleophilicity of the L and X ligands based on the k_3/k_2 values. Analyze this ordering on the basis of the Hard–Soft theory of acids and bases and other factors thought to affect nucleophilicity.

(c) Calculate k_4 from the overall equilibrium constant and the kinetic data for the three systems with X = MeIm and L = PPh$_3$, P(OPr)$_3$, and py. Are these values of k_4 consistent with the mechanistic proposal? Suggest how the calculated value of log K was obtained.

(d) Calculate k_4 for the system with L = PPh$_3$ and X = py. Is the result consistent with other results in the table?

2. Will the following complexes be labile or inert as defined by Taube? Cr^{II}(L)$_6$, octahedral low spin; V^{III}(L)$_6$, octahedral; Rh^{II}(L)$_6$, octahedral.

3. Various types of evidence indicate that the substitution reactions on Co(OH$_2$)$_6^{2+}$ in water have a **D** mechanism. If this is correct, explain why the reaction rates are observed to be first-order in the entering ligand.

4. (a) Use the *d* orbital energies in Table 3.14 to calculate the crystal field activation energies for an octahedral *d*^{3} system undergoing substitution by a trigonal pyramidal and a pentagonal bipyramidal

transition state. Use these results and the data in Table 3.15 to predict the most favorable transition state of regular geometry and the mechanism of substitution for such a system.

(b) Calculate the crystal field activation energy for a square planar d^8 system undergoing associative substitution by trigonal bipyramidal and square pyramidal transition states.

5. The reaction of aqueous vanadium(III) with salicylic acid has the following rate law.

$$\text{Rate} = \left(6.0 + \frac{2.6}{[H^+]} \right) [V(III)]_{\text{total}} \, [\text{salicylic acid}]_{\text{total}}$$

The relevant ionization reactions in this system are the following:

$$V(OH_2)_6^{3+} \underset{1.4 \times 10^{-3} \text{ M}}{\overset{K_m =}{\rightleftharpoons}} V(OH_2)_5(OH)^{2+} + H^+$$

$$C_6H_4(OH)(CO_2H) \underset{1.6 \times 10^{-3} \text{ M}}{\overset{K_{a1} =}{\rightleftharpoons}} C_6H_4(OH)(CO_2)^- + H^+$$

$$C_6H_4(OH)(CO_2)^- \underset{1 \times 10^{-12} \text{ M}}{\overset{K_{a2} =}{\rightleftharpoons}} C_6H_4(O)(CO_2)^{2-} + H^+$$

(a) Propose two reaction schemes that are consistent with the rate law.

(b) Calculate the specific rate constants for each scheme in (a) and determine which, if any, is more probable on the basis of other data in Table 3.23.

CHAPTER 4

1. For the following complex, draw diagrams describing a process that :

(a) Causes racemization but no isomerization.

(b) Causes cis–trans isomerization.

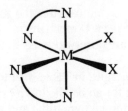

2. For each of the isomers of the following asymmetric chelate, predict the products if the reaction proceeds by one-ended dissociation of A' to give a trigonal bipyramidal intermediate with the monodentate ligand in the trigonal plane. Assume that ring closing occurs with equal probability along the trigonal edges. Indicate if a new structural isomer or stereo isomer of the reactant has formed in each case.

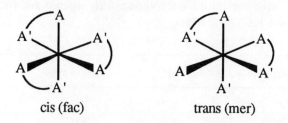

cis (fac)   trans (mer)

3. Stereochemical mobility in a square-based pyramidal molecule may be due to an "anti-Berry" pseudorotation in which the intermediate in the normal pseudorotation is actually the reactant. Sketch an anti-Berry process for a square-based pyramidal molecule.

4. Show with clear diagrams why an η^2-C_6H_6 system should be more readily fluxional than an η^4-C_6H_6 system.

CHAPTER 5

1. Explain the trends in the rate constants for the following reaction.

$$Cr(CO)_5L + *CO \longrightarrow Cr(*CO)(CO)_4L + CO$$

Compound	k (s^{-1}, 30°C)[a]
$Cr(CO)_6$	1.0×10^{-12}
$Cr(CO)_5(PMe_2Ph)$	1.5×10^{-10}
$Cr(CO)_5(PPh_3)$	3.0×10^{-10}
$Cr(CO)_5Br^-$	2.0×10^{-5}
$Cr(CO)_5Cl^-$	1.5×10^{-4}

[a] Atwood, J. D.; Brown, T. L. *J. Am. Chem. Soc.*
1976, *98*, 3160, and references therein.

2. For the following reaction, the rate constant in toluene has been found to be independent of the nature and concentration of L.

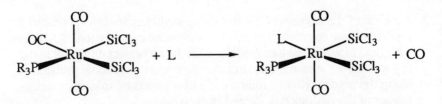

The kinetics have been studied for various PR_3 ligands and the rate constants are given in the following table.

Phosphine	pK$_a$	Cone Angle	$-\log k$ [a]
P(OCH$_2$)$_3$CEt	1.74	101	5.96
P(OMe)$_3$	2.6	107	5.14
PPh(OMe)$_2$	2.64	120	4.95
PPhMe$_2$	6.50	122	4.72
P(OPh)$_3$	−2.0	128	4.70
PPh$_2$(OMe)	2.69	132	4.64
PPh$_2$Me	4.57	136	4.15
P(p-ClC$_6$H$_4$)$_3$	1.03	145	3.17
P(p-FC$_6$H$_4$)$_3$	1.97	145	3.12
PPh$_3$	2.73	145	3.05
P(p-MeC$_6$H$_4$)$_3$	3.84	145	3.08
P(m-MeC$_6$H$_4$)$_3$	3.3	165	2.89
P(m-ClC$_6$H$_4$)$_3$	1.03	165	3.02

[a] Chalk, K. L.; Pomeroy, R. K. *Inorg. Chem.* **1984**, *23*, 444; at 40°C in toluene, k in s^{-1}.

(a) Suggest the reaction mechanism based on the kinetic observations with varying L.

(b) The relative constancy of the rate constants for the series of PR$_3$ with a cone angle of 145° but different pK$_a$ indicates that the reaction is relatively insensitive to σ-bonding effects. However, the rate constant seems to vary with the cone angle. Make a plot of log k versus cone angle and analyze the steric effects in terms of their influence on the ground state and transition state for the reaction. (*Hint*: See Goering et al. *Organometallics* **1987**, *6*, 650.)

3. Nineteen-electron organometallic complexes are rare and generally quite reactive. Recently, the ligand substitution kinetics have been studied for the following system.

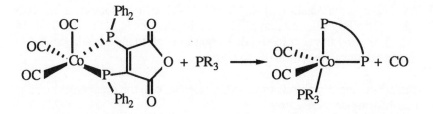

Although formally 19-electron systems, these complexes may be viewed as Co(I) with a reduced ligand radical (P_2^-). In any case, the mechanism of substitution is of interest and the following kinetic data have been obtained. All the data are for 1.4×10^{-3} M Co(CO)$_3$(P-P).

Table A. Rate Constants (25°C) for Co(CO)$_3$(P-P) + PPh$_3$ in CH$_2$Cl$_2$

[PPh$_3$] (M)	0.148	0.296	0.371	0.445	0.556	0.704
$10^3 \times k$ (s^{-1})	5.48	5.45	5.47	5.52	5.42	5.50

Table B. Rate Constants for Co(CO)$_3$(P-P) + PPh$_3$ in CH$_2$Cl$_2$

t (°C)	10.0	15.0	20.0	25.0	30.0
$10^3 \times k$ (s^{-1})	0.598	1.26	2.67	5.47	10.2

Table C. Rate Constants (25°C) for Co(CO)$_3$(P-P) + PR$_3$ in CH$_2$Cl$_2$

PR$_3$	PPh$_3$	P(OPh)$_3$	PMePh$_2$	PBu$_3$	P(OMe)$_3$
$10^3 \times k$ (s^{-1})	5.47	5.18	5.20	5.58	5.03

(a) Are the kinetic results consistent with an I_a mechanism?

(b) Are the kinetic results consistent with an A mechanism?

(c) Are the kinetic results consistent with rate-controlling chelate ring opening followed by substitution? (See Scheme 5.2.)

(d) Are the kinetic results and activation parameters for PPh$_3$ consistent with an I_d mechanism?

4. It has been observed that *cis*-(Et$_3$P)$_2$Pt(H)(Solv)$^+$ (Solv = methanol or acetone) reacts with 1-hexene but only brings about the isomerization of the 1-hexene to 2-hexene, and 3-hexene. Suggest a mechanism for this process.

5. Assume that the reaction of Cp*Rh(CO)(Kr) with cyclohexane involves rate-controlling dissociation of Kr. Verify that the k_{exp} has the form of Eq. (5.39) and the conclusion of Bergman et al. that α should be independent of the nature of the alkane.

6. The following reaction is of possible relevance to the mechanism of the hydroformylation reaction.

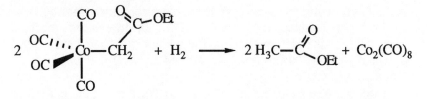

The dependence of the reaction rate on H_2 and CO concentrations has been determined at 35°C in *n*-heptane and the results are given in the following table.

$10^3 \times [H_2]$ (M)	3.68	3.68	9.94	3.30	2.59	2.13
$10^3 \times [CO]$ (M)	1.31	1.31	3.56	2.30	4.16	5.35
$10^5 \times k_{exp}$ (s^{-1})	1.89	1.93	1.60	1.03	0.43	0.25

(a) Deduce the rate law that seems to describe the results best.

(b) Suggest a mechanism that is consistent with the rate law from (a).

(c) Suggest a further type of experiment or kinetic study that would help to confirm your suggestion in (b).

CHAPTER 6

1. The oxidation of iron(II) by lead(IV) might proceed by either of the following mechanisms.

     Mechanism A   Mechanism B

$Fe^{II} + Pb^{IV} \rightleftharpoons Fe^{III} + Pb^{III}$ $Fe^{II} + Pb^{IV} \rightleftharpoons Fe^{IV} + Pb^{II}$

$Fe^{II} + Pb^{III} \longrightarrow Fe^{III} + Pb^{II}$ $Fe^{II} + Fe^{IV} \longrightarrow 2\,Fe^{III}$

(a) Assume a steady state for Pb^{III} and Fe^{IV} and develop the rate law for each mechanism.

(b) Could these alternatives be distinguished experimentally?

2. Use the data in Table 6.2 and the Marcus relationship, Eq. (6.33), to predict the rate constants for the following reactions.

(a) $Co(sep)^{3+} + Cr(OH_2)_6^{2+} \longrightarrow Co(sep)^{2+} + Cr(OH_2)_6^{3+}$

(b) $Fe(OH_2)_6^{3+} + Ru(NH_3)_6^{2+} \longrightarrow Fe(OH_2)_6^{2+} + Ru(NH_3)_6^{3+}$

3. For the following reaction, $k = 4.2 \times 10^2 \ M^{-1} \ s^{-1}$ and $K = 1.3 \ (25°C)$.

$$V(OH_2)_6^{3+} + Cr(bpy)_3^{2+} \longrightarrow V(OH_2)_6^{2+} + Cr(bpy)_3^{3+}$$

Use additional data in Table 6.2 to estimate the self-exchange rate constant for $Cr(bpy)_3^{2/3+}$.

4. Predict the products of the following reactions.

(a) $Cr(OH_2)_6^{2+} + (NH_3)_5Co(CN)^{2+} \longrightarrow$

(b) $Ru(OH_2)_6^{2+} + (NH_3)_5CoCl^{2+} \longrightarrow$

CHAPTER 7

1. In Figure 7.1, assume that the photoactive state A can react by two pathways to give products P_1 and P_2 with rate constants k_1 and k_2, respectively. Develop the expression for the quantum yields of the two products.

2. The energy levels for the d–d transitions in an octahedral chromium(III) complex are given in Figure 7.3. Suggest two explanations for the observation that the quantum yield for photosubstitution on such complexes is independent of the irradiation wavelength over the region 350- to 700-nm.

3. Use Adamson's rules to

(a) Predict the product of the photolysis of *cis*-Cr(NH$_3$)$_4$(Cl)$_2^+$.

(b) Explain the very low quantum yield for photoaquation of *trans*-Cr(en)$_2$(OH$_2$)$_2^{3+}$.

4. For Scheme 7.11, develop the kinetic expression that can be used to determine the various rate constants from the dependence of the rate on [CO] and [THF].

5. The activation volumes (cm^3 M^{-1}) for the photolysis of a chromium(III) and a rhodium(III) complex are given in the following table. Suggest rationalizations for the similarities and differences.

	$\Delta V^*(\Phi_{Cl})$	$\Delta V^*(\Phi_{NH_3})$
$Cr(NH_3)_5Cl^{2+}$	-13	-9.4
$Rh(NH_3)_5Cl^{2+}$	-8.6	$+9.3$

CHAPTER 8

1. Describe in detail the complete mechanistic sequence proposed by Finke and co-workers for the reaction of diol dehydrase with ethylene glycol.

2. (a) For a system showing competitive inhibition, develop expressions for the slope, intercept when $[S]^{-1} = 0$, and intercept when $v_i^{-1} = 0$ for Lineweaver–Burke plots, by rearrangement of Eq. (8.16).

   (b) For a system showing uncompetitive inhibition, develop expressions for the slope, intercept when $[S]^{-1} = 0$, and intercept when $v_i^{-1} = 0$ for Lineweaver–Burke plots, by rearrangement of Eq. (8.18).

   (c) Draw sketches to show how the Lineweaver–Burke plots would vary with [I] for competitive and uncompetitive inhibition.

3. Assume that the hydrolysis of phenyl acetate by carbonic anhydrase proceeds by a mechanism analogous to that for CO_2 hydration. Describe the sequence of reactions that would be involved in the ester hydrolysis.

4. Show that the steady-state solution for Scheme 8.5 yields Eq. (8.28).

5. Consider the electronic structure of the reactants and suggest why the rate-controlling step for CO binding to myoglobin may be the Fe—CO bond formation, while this step is not rate-controlling for O_2 binding.

Chemical Abbreviations

acac	acetylacetonate; 2,4-pentanedioate
azacaphen	1-methyl-3,13,16-trithia-6,8,10,19-tetraazabicyclo[6.6.6]eicosane
bpy	2,2'-bipyridyl
Bu	butyl
Cp	cyclopentadienide
Cp*	pentamethylcyclopentadienide
Cy	cyclohexyl
cyclam	1,4,8,11-tetraazacyclotetradecane
dien	diethylenetriamine
DMF	N,N-dimethylformamide
DMSO	dimethylsulfoxide
EDTA	ethylenediaminetetraacetate
en	ethylenediamine; 1,2-diaminoethane
Et	ethyl
hfac	hexafluoroacetylacetonate
isn	isonicotinamide
Me	methyl
nic	nicotinamide
ox	oxalate
Pf$_3$	trifluoromethyl pyrazolyl
Ph	phenyl
phen	1,10-phenanthroline
Pr	propyl
py	pyridine
tacn	1,4,7 triazacyclononane
THF	tetrahydrofuran
tn	trimethylenediamine; 1,3-diaminopropane
TPP	tetraphenylporphinate
t-tacn	1,4,7-trithiacyclononane
sar	sarcophagine; 3,6,10,13,16,19-hexaazabicyclo[6.6.6]eicosane
sep	sepulchrate; 1,3,6,8,10,13,16,19-octaazabicyclo[6.6.6]eicosane

Index